ECM/87

Educational Computing in Mathematics

ECM/87

Educational Computing in Mathematics

Proceedings of the International Congress on
Educational Computing in Mathematics
Rome, Italy, 4–6 June, 1987

Edited by

T. F. BANCHOFF
Department of Mathematics
Brown University
Providence, RI, U.S.A.

M. EMMER
Department of Mathematics
University of Rome 'La Sapienza'
Rome, Italy

I. CAPUZZO DOLCETTA
Department of Mathematics
University of Rome 'La Sapienza'
Rome, Italy

H. KOÇAK
Lefschetz Center
    for Dynamical Systems
Brown University
Providence, RI, U.S.A.

M. DECHAMPS
Department of Mathematics
University of Paris-Sud
Orsay, France

D. L. SALINGER
Department of Pure Mathematics
University of Leeds
Leeds, United Kingdom

1988

NORTH-HOLLAND
AMSTERDAM ● NEW YORK ● OXFORD ● TOKYO

ISBN: 0 444 70456 6

Publishers:
ELSEVIER SCIENCE PUBLISHERS B.V.
P. O. Box 1991
1000 BZ Amsterdam
The Netherlands

Sole distributors for the U.S.A. and Canada:
ELSEVIER SCIENCE PUBLISHING COMPANY, INC.
52 Vanderbilt Avenue
New York, N.Y. 10017
U.S.A.

PRINTED IN THE NETHERLANDS

INTRODUCTION

This volume contains contributions to the International Conference ECM/87 "Educational Computing in Mathematics" held at the University of Rome "La Sapienza", June 4-6, 1987.

The idea of the conference was to share experiences on computer aided teaching and also to point out some topics where current mathematical research takes advantage of the facilities offered by a computer. This idea originated in 1986 among the members of the EEC (European Economic Community) project "Utilisation des moyens informatiques dans l'enseignement des mathématiques" and the CNR (National Council of Research) supported "Progetto strategico: nuove tecnologie per la didattica". About 120 participants from various European countries and the U.S.A. attended the Conference which was organized in invited and contributed talks, software demonstration sessions and a symposium on "The Italian situation in computer aided teaching" coordinated by M. Falcone, with the interventions of C. Bohm, C. Pucci, and V. Villani, President of the Unione Matematica Italiana.

These proceedings contain the following three sections: Mathematical Experiments, Teaching Experiences, and Software Reports, covering all invited talks as well as a selection of contributed papers and software demonstration reports.

ACKNOWLEDGEMENTS

As organizers of the congress we would like to take this opportunity of thanking all those who helped in making the congress and the software exhibition successful.

First of all Michele Emmer's students of Advanced Calculus I from the Department of Physics, without whose assistance the congress would never have been possible: Massimiliano Bufacchi, Gaspare Lo Curto, Alessandro Mattacchini, Silvia Mazzei, Simona Mei, Vincenzo Monaco, Pierfrancesco Moretti, Gianluca Morgante, Salvatore Motta, Andrea Nigrelli, Giovanni Oliva, Dina Onali, Ernesto Palumbo, Sabrina Paoliani, Mario Paolucci, Maria Alessandra Papa, Andrea Parri, Gherardo Piacitelli, Nadia Piraccini, Alberto Polimanti, Emanuela Pompei, Maria Paola Pompilio, Paola Puppo, Indiana Raffaelli, Valeria Ricci, Walter Rinaldi, Massimiliano Roccetti, Stefano Romagnoli, Barbara Romano, Valeria Romano Franchi, Claudio Ruggeri, Stefano Sarti, Francesca Sarto, Michele Schillizzi, Bruno Sciarretta, Barbara Sciascia, Andrea Sermoneta, Francesca Spacocci, Sandra Stellato, Antonella Tajani, Anna Taverna, Gianlorenzo Tentori, Andrea Tesseri, Marco Tomini, Antonella Tufano, Sara Velardi, Claudia Venditti, Roberta Venditti, Lorenzo Violante, Grazia Flavia Vistoso, Aldo Winkler, Andrea Zanela, Massimiliano Zanot.

A particular word of thanks to the secretaries of the Department of Mathematics Mrs. Caporusso, Falco and Norrito, and to all the personnel, in particular Mr. Grossi; the company TELAV for the equipment of the Aula III of the Department (its technicians were able to set up the video projector just the night before the conference); our colleague G. Ferrarese for the 16 mm. projector; the companies IBM, Apple, BBC Acorn for the exhibition.

The reception was offered by the Company Olivetti; the same Company has helped the Organizers in setting up the Workshop and other technical facilities.

Special thanks to the Company Natalizi for the luxurious coffee-breaks; all the colleagues in the Department, even not interested in the congress, were really attracted by the morning receptions.

The financial support for the congress was provided by the funds of the University of Rome "La Sapienza", the CNR, the Company OLIVETTI, and by the funds of the EEC.

The members of the organizing committee were: D. Knapp (University of Leeds), F. Cottet Emard (University of Paris Sud), I. Capuzzo Dolcetta, M. Emmer, M. Falcone, S. Finzi Vita, L. Loreto, M. Picardello (University of Rome "La Sapienza").

CONTENTS

2. TEACHING EXPERIMENTS 125

1. MATHEMATICAL EXPERIMENTS

ECM/87 - Educational Computing in Mathematics
T.F. Banchoff et al. (editors)
© Elsevier Science Publishers B.V. (North-Holland), 1988

COMPUTER ENHANCED LEARNING:
THE LOUGHBOROUGH EXPERIENCE

A. C. BAJPAI and P. A. CARRUTHERS

Department of Engineering Mathematics, University of Technology,
Loughborough, Leics LE11 3TU, England

The use of microcomputers has, over the years, become a professionally acceptable practice at all levels of mathematics education. Of these levels the slowest to respond, for one reason or another, has been higher education. This paper outlines our attempts at Loughborough to combat this problem by producing flexible and easy-to-use software units, and describes our first steps towards introducing them into our teaching activities. Our vision of Computer Enhanced Learning is also put forward in connection with these units.

1. BACKGROUND – THE MIME[1] PROJECT

Our enthusiasm and expertise at Loughborough for producing educational software has evolved over a four year period which started with the ambitious MIME Project. The aim of the project was to produce software units on Mechanics topics for Advanced level (16–18 years old) and first year university students. Our hope was to move away from the traditional approach of a programmed text and to utilise more fully the power and flexibility of the micro. This meant that the software should allow the greatest possible degree of interaction to encourage the user to learn through his own experimentation or to allow a teacher to present his class with an example or situation of his own choosing. With such an emphasis on user participation it was necessary for the units to be extremely user-friendly and robust bearing in mind the possible inexperience of potential users.

A fuller account of the MIME Project is given in a series of three papers [1], [2] and [3] but to summarise the objectives of the MIME units, they should:

(a) aid understanding
(b) add interest
(c) be interactive
(d) be user-friendly
(e) be robust
(f) enhance presentation.

Figure 1 is taken from one of the units – Relative Motion.

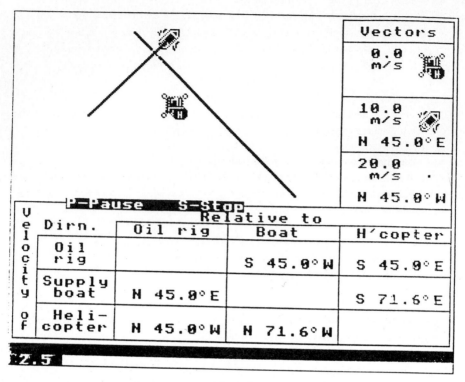

Figure 1

2. COMPUTER ENHANCED LEARNING

Point (f) brings to light the expression *'Computer Enhanced Learning'* or CEL which was first coined by the first author in order to help describe the type of software that we wished to produce. Despite a plethora of acronyms connecting computers with education (CAL[2], CBL[3], CIT[4], CML[5], CAI[6]), none completely fulfilled our vision of what such software should be.

Much of the software we looked at (including early MIME units) comprised the following elements:

(i) Introductory text

(ii) Fixed examples

(iii) Theory

(iv) Interactive section.

We felt that introductory text and theory wasted the micro's power and would better serve as accompanying written material. Fixed examples could be incorporated into the interactive section by supplying the user with the appropriate parameter values. In this way the software would be more of the simulation type than the step-by-step type. The emphasis is then placed on the user rather than the micro – the lecturer or tutor using the machine as a tool to enhance his presentation, or the student using the micro to conduct guided experiments through discovery.

When dealing with undergraduate students, a higher level of self-motivation is to be expected. Consequently our CEL software for such students need be less reliant on gimmicks such as showy animation or touches of humour to provide incentive to learn. It was found that the advanced programming techniques used in MIME animations could be put to use in writing undergraduate CEL units but in a more subtle way, creating greater clarity through selective highlighting, for example, or providing a rolling menu to fit in a confined screen area.

3. THE CEL UNITS

The CEL units that have so far been produced at Loughborough were obviously written in the light of our own teaching experience, and with the specific needs of our own students in mind. However in accordance with the CEL philosophy, they were made flexible enough to be of interest to any lecturer or tutor involved in those particular topics.

Many of the lessons we had learnt from the MIME Project were borne in mind when producing these units, particularly the importance of standardisation. It had been found that standardisation helps:

(a) The user;
Methods of input and other interaction need only be learnt once. The general style and appearance of screen display changes little from one program to another, and indeed from unit to unit.
(b) The programmer;
Common subroutines need only be written once. A rigid format leaves the programmer little to think about in terms of program specification.

The net effect of standardisation is to save time for both parties.

One new standard that was adopted for the CEL units was the facility for 5-key input. Five keys, the four cursor keys, and RETURN could be found either on the keyboard or on a remote input device, and could be used to drive any of our programs by using the cursors to increment or decrement parameter values, or to move a highlighting bar or to choose initial conditions, etc, and using RETURN to confirm each input. This standard was easy for students and teachers to adapt to, as the method of selection was nearly always obvious, although it did add considerably to programming complexity.

Units so far completed are:
          Complex Transformations
          Poles and Residues
          Numerical Integration
          Numerical Solutions of Equations
          Numerical Solution of Non-Linear Equations
          The Water Tank
          Numerical Solution of Ordinary Differential Equations
          Cubic Splines

(In addition, 13 CEL units on Mechanics and 5 units on Statistics have been published by John Wiley & Sons Ltd, Chichester, England.)

4.   TESTING AND EVALUATION

The process of testing and evaluating any educational software is one which, in order to obtain true reflection, must be carried out over a number of years. Our new CEL units have so far received only one academic year of classroom testing. This process will carry on in the future to create a clearer picture of their true worth, but meanwhile some feedback has been forthcoming from our own students, and many of these comments have instigated revisions to the software. For example, a group of first year Mathematical Engineers were given a computer based tutorial on Numerical Integration, having previously been taught the subject in a conventional lecture. The tutorial was meant to 'enhance' the students' knowledge – to take the hard facts and information which had already been taught, and turn it into an intuitive feeling for the Trapezium and Simpson's Rules. In practice, however, the tutorial was found to add little to the students' grasp of the subject. What they wanted was something to help them understand about the errors connected with these two methods. A feature was subsequently added to depict the error over each strip giving, over the whole interval, a step function *looking* like $f''(x)$ ór $f^{(iv)}(x)$ (Figure 2). Subsequent trials showed that this feature gave the tutor more power and authority when demonstrating the topic, and instilled less fear into students already somewhat confused by complicated error formulae.

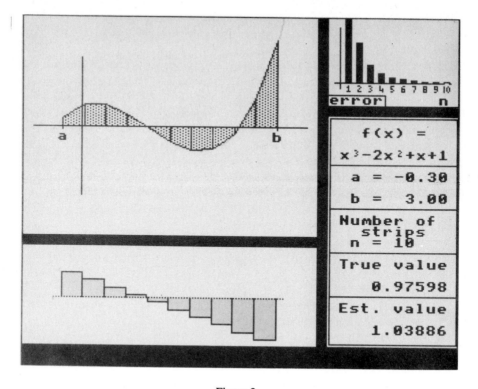

Figure 2

As mentioned above, animation was not considered an important factor for the CEL units. This fact was highlighted when the 'Water Tank' package (Figure 3) was used with some first year Electrical Engineers. The animated tank on the left of the screen proved to be something of a distraction for these students – they had all had programming exposure as part of the course and were curious to discover "How was it done?" While this took their attention away from the purpose of the software it must be said that the interest and motivation of the class was boosted.

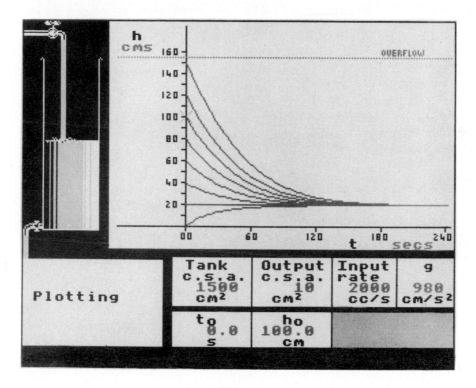

Figure 3

Overall, students felt that they benefitted from interactive learning situations and were eager to experiment with software or to be guided through important examples either by tutorial sheets or by the tutor himself.

5. CONFERENCE REACTION

The reaction of conference delegates to our, albeit rather short, software demonstration was both gratifying and encouraging. We were pleased to note that many delegates were keen to point out the possible applicability of individual pieces of our software to their own teaching needs. This tends to confirm our belief in the importance of flexibility. Particular interest was paid to the package on Complex Transformations (Figure 4), a subject whose abstract nature often creates teaching difficulties.

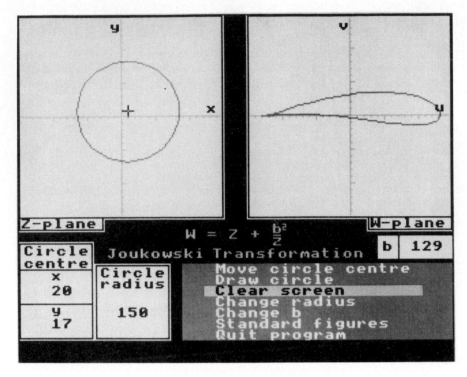

Figure 4

The main dissatisfaction to be expressed was with our choice of host machine – Acorn's BBC microcomputer. The reason for this choice was that the cheapness, robustness and ease of programming of the BBC had made it the standard for educational establishments in Great Britain. This prevalence, however, stops at our Island's shores, at which point the IBM PC takes over as the natural choice. Whilst we would very much like to see our software available for the IBM, the cost, unfortunately, is prohibitive. A feasibility study of the conversion of one MIME unit (Momentum and Impacts) to run on the IBM proved to take as long to write as did the original. The funding available for the production of educational software is restricted enough, without the added burden of having to write several versions of each unit for different micros.

6.   CONCLUSIONS

Our experience has shown that software production is a very costly business, and with solid investment in it being so slack, we must all beware the danger of wasting money by recreating existing software or 'reinventing the wheel'. Many mathematical subjects (particularly numerical ones) are natural targets for educational software, but in many cases this software already exists; a more beneficial target would be those subjects whose nature is more difficult for students to understand and whose consequent applicability to the micro is less easy to visualise.

The implementation of such a globally cost-effective scheme calls for greater cooperation and communication between participating bodies, greater monetary assistance, and greater involvement and thoughtfulness from hardware manufacturers. Let us all hope it can be achieved.

FOOTNOTES

1   Micros In Mathematics Education
   (Founded in 1983 under the Directorship of the first author)
2   Computer Assisted Learning
3   Computer Based Learning
4   Computer Illustrated Text
5   Computer Managed Learning
6   Computer Aided Instruction

REFERENCES

[1]   Bajpai, A.C., Fairley, J.A., Harrison, M.C., Mustoe, L.R., Walker, D. and Whitfield, A.H., The MIME Project at Loughborough – A first report, in: Int. J. Math. Educ. Sci. Technol., **15** (1984), 6, 781–810

[2]   Bajpai, A.C., Fairley, J.A., Harrison, M.C., Mustoe, L.R., Walker, D. and Whitfield, A.H., Mathematics and the micro: some hints on software development, in: Int. J. Math. Educ. Sci. Technol., **16** (1985), 3, 407–412

[3]   Bajpai, A.C., Fairley, J.A., Harrison, M.C., Mustoe, L.R., Walker, D. and Whitfield, A.H., The MIME Project at Loughborough – A second report, in: Int. J. Math. Educ. Sci. Technol., **18** (1987), 2, 301–313

ECM/87 - Educational Computing in Mathematics
T.F. Banchoff et al. (editors)
© Elsevier Science Publishers B.V. (North-Holland), 1988

EDGE:THE EDUCATIONAL DIFFERENTIAL GEOMETRY ENVIRONMENT

Thomas Banchoff,Richard Schwartz

Mathematics Department,Brown University

Providence,RI 02912,USA

Over the past eighteen years,courses in differential geometry at Brown University have been using computer-generated films,slides and videotapes for presenting geometric phenomena. Most of the time these materials were used in conjunction with lectures,for illustrating theorems and providing motivation for new topics. In recent years,however,the involvement of students in differential geometry has intensified due to the development of interactive computer graphics programs which allow students to take a direct part in investigation of curves and surfaces. Three times over the past year students in the course in elementary differential geometry have partecipated in weekly one-hour laboratory sessions as a supplement to the ordinary classroom lectures. The laboratory is equipped with enough Apollo workstations to allow each students to work with a single machine. The aim of this paper is to report on these experiences.

1. INTRODUCTION

For twelve years beginning in the late 1960's, the first-named author worked primarily with Charles Strauss, who had recently completed a Ph.D thesis in interactive three-dimensional design. The products of that collaboration have been described in several articles,[1], [2], [3], culminating in an invited address at the International Congress of Mathematicians in Helsinki in 1978 [4]. Commencing in 1981, the collaboration has been mainly with students. David Salesin, Steven Feiner, and the first named author developed a high-level animation language which was particularly well suited for presentation and analysis of surfaces in three- and four-dimensional spaces [5]. Simultaneously, Timothy Kay, Edward Grove, and Richard Hawkes, three undergraduate students in the graduate graphics course taught by Professor Andries

van Dam, developed software for the first interactive programs
to be used in a differential geometry laboratory. Hawkes continued as an assistant in that course for two semesters, adding greatly to the capabilities of DCP, the "Differential Curves Package"
in response to the classroom activities. Experience with this
program pointed out numerous limitations in its structure, in
particular in the methods for defining new functions and for ma-
nipulating them by rotations, scaling, and changes in parameters.
In the meantime the techniques of developing user-friendly interfaces had progressed significantly, particularly in the laboratory program CAMP, the "Complex Analysis Mapping Package", developed in the graduate graphics course by juniors Herbert "Trey"
Matteson and Richard Schwartz for use in the undergraduate complex variables course. Two years ago the differential geometry
program took on an entirely new form through the efforts of Robert Shapire and Kathleen Curry. Their final project in the graphics course made striking improvements in earlier approaches
and produced EDGE, the "Educational Differential Geometry Environment". During the following summer, Shapire and Curry worked
together with Schwartz to prepare the enhanced program for use
in the Fall semester course in differential geometry. Schwartz
acted as assistant in that course, and prepared a final report
as his senior project in the combined Mathematics-Computer Science Bachelor of Science degree. The following pages are largely
taken from his description of this ongoing project. The first-
named author would like to thank him and all the other students
who have contributed to the development of these programs through
their participation in the graphics course and the differential
geometry courses. He would also like to acknowledge the cooperation and support of Professor van Dam and the members of the
graphics research group at Brown University, and the financial
support provided by Dean Harriet Sheridan and Provost Maurice
Glicksman at Brown University, and several grants from the
National Science Foundation.

2. EDGE – EDUCATIONAL DIFFERENTIAL GEOMETRY ENVIRONMENT

Elementary differential geometry, the study of curves and surfaces in the plane and in 3-space, combines the abstractions of
calculus with the visual representations of geometry. Because
of the visual nature of the subject, concrete models or diagrams
can greatly simplify the learning experience. In the past,
teachers have found it very difficult to provide enough illustrations to enhance the students' understanding of the subject
matter. One can spend hours trying to copy large and detailed
diagrams into chalkboards, or constructing complicated models
out of cardboard, plaster, or wire, and still provide students
with only a minimum of effective visual information.

However, recent developments in computer graphics, and in the
use of the computer as a classroom tool in particular, have
provided a fast and powerful way for teachers to present information, and for the students to experiment on their own with rather
complex concepts.

EDGE--Educational Differential Geometry Environment--is an
interactive software package developed for the differential geometry course at Brown University. In the Fall semester of 1986,
in addition to the normal course hours, the class met once a week
for an hour in the interactive computer graphics laboratory,
under the guidance of the authors. Each week, students were presented with particular concepts to investigate and problems to
solve. They were encouraged to do independent work and to participate in group efforts. In addition, students were assigned a
term project on some area of differential geometry, and several
students used EDGE to produce images for their final reports.

EDGE is an educational tool which helps to illustrate
the concepts and principles of differential geometry. Specifically, it allows the student to examine the geometry of curves
and surfaces in two- and three-space. (For convenience, we
will only consider curves, although most of the discussion
applies to surfaces as well.) The student indicates the curve
to be examined by defining its coordinate functions parametri-

cally--by entering the three functions X(t), Y(t) and Z(t). The student also needs to tell the computer the range for the function variable,t,plus the number of equally-spaced points at which to evaluate the curve. The computer will perform all necessary calculations and display the curve on the screen. Once generated, the object can be manipulated in a number of ways--rotated, translated, animated, projected, etc.--so that the student can get a full appreciation of what the object "looks like". Additionally, the computer can generate at each point a number of vectors with which differential geometers are concerned--such as velocity and acceleration--as well as the curvature and torsion functions. Using other features of the package--tubes, generalized vector functions, and multipli windows--students can gain insight into the ways these vectors and functions relate to the curve.

In this paper, we will first review some of the basic mathematics used in the differential geometry course. Next, we will show some of the mathematical concepts which EDGE has been used to illustrate, many of which are very difficult, if not impossible, to treat without the use of the computer. We will thendiscuss some of the special features of the program in more detail.

3. BASIC TERMINOLOGY OF DIFFERENTIAL GEOMETRY

A curve in 3-space is defined by three coordinate functions X(t), Y(t), and Z(t), of a single variable, t. For example, a helix, one of the simplest three-dimensional curves, is defined as:

$$X(t) = \cos(t) , Y(t) = \sin(t) , Z(t) = t ,$$

or

$$X(t) = \{\cos(t) , \sin(t) , t \} .$$

The variable t can be thought of as time, and curves in general can be thought of as the paths traced out by particles moving in space.

As one might imagine, the first thing a differential geometer

ve in the direction of its principal normal produces a curve which has linking number 1 with the original curve. This can be shown by creating the additional curve

$$\{X(t) + e\ P(t)\},$$

where e is a small number, and displaying both curves using the Hidden Edge feature to show over- and under-crossings.

6.COMMUNICATION FEATURES

We note at this point how the "hands on" interactive nature of EDGE greatly influences the types of questions which can be asked. In a typical class, the professor' would propose a topic of inquiry, such as examining the torsion zeros of a space curve, and it would be up to the student to raise and examine issues. Using any of the methods described above, in a single class period, a student can examine properties of a large number of curves and curve families. He or she can generate conjectures on the basis of these observations, then try to use analytical techniques to prove some of these conjectures, and then come back and share those insights with the class. This process happens much more readily in the setting of an interactive computer graphics laboratory.

EDGE has a number of special features which allow the students to communicate with the professor and with other students. One such feature is the *library* system. There are certain curves which come up throughout the semester. These are saved in a library of curves (and tubes and surfaces). A curve is saved along with the definitions of $X(t)$, $Y(t)$ and $Z(t)$, the interval on which the variable t is defined, the number of iterations for t, and default values for all the coefficients. There is a *global* library where the professor and assistant may save examples for the entire class, and each student has a personal account with a *local* library where he or she can save curves for future study.

Furthermore the *view-saving* feature enables a user to regenerate a particular view, along with the definitions of curves, tubes, and generalized vectors, the values of the axes in

wants to do with an equation is to differentiate it. As in physics, the first derivative of a curve yields the *velocity* vector, and the second derivative gives the *acceleration*. The program automatically computes the first three derivatives symbolically.

These three derivatives are orthonormalized to give the *Frenet frame* consisting of the *unit tangent* T, the *principal normal* P, and the *binormal* B. At a point where the first and second derivatives are non-zero, the unit tangent is the unit vector in the direction of the cross-product of the first two derivatives and the principal normal is the cross-product of the binormal and the tangent.

Differential geometers are interested primarily in two basic functions defined at all points of the curve, the curvature, and the torsion. The *curvature* measures the degree to which the curve is "turning", or moving away from a straight line and the torsion. The *torsion* measures the rate at which a curve is twisting away from a plane at a given point. The program computes these quantities by the formulas

$$ k(t) = \frac{\|X' \times X''\|}{\|X'\|^3} , \quad \tau(t) = \frac{X''' \cdot X' \times X''}{\|X' \times X''\|^2} . $$

4. THE EDGE ENVIRONMENT

The EDGE environment is an Apollo workstation, including a black-and-white graphics monitor with moderately high resolution, a keyboard to enter date, and a *"mouse"*. EDGE is a menu-driven interactive graphics package. *"Interactive"* refers to the fact that the user has complete control over what happens in a given session. *"Menu-driven"* refers to the way in which the user interacts with EDGE. The user tells EDGE what to do by selecting an item from one of the various pull-down menus. Menu selections are made by placing the mouse on the selection and "clicking" on one of the buttons.

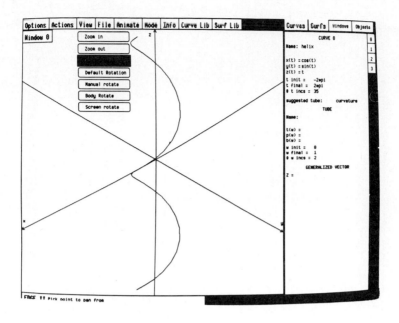

Figure 1

Selecting an action from one of the menus.

Some of the menu choices require no further input beyond selection (for example, the *"draw"* command will immediately attempt to
draw the currently defined curve on the screen). However,
many of the choices require more information from the user
(such as any command selection which requires keyboard input).
This latter cases are handled by what are called "dialogue
boxes". Dialogue boxes provide a way for the user to communicate
with EDGE. There are three different types of activities which
occur through dialogue boxes. First, a user may be prompted
to enter data from the keyboard, either alpha-numeric (e.g.the
definition of a curve) or numerical (e.g.the endpoints of
an interval).

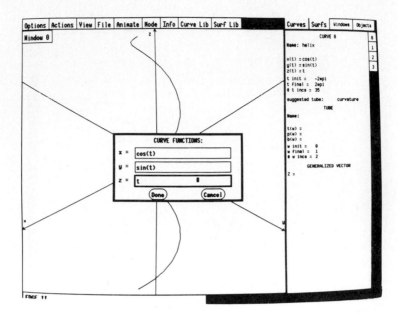

Figure 2

Entering curve definitions in a dialogue box.

Second, a user may be given a list from which to make a selection, such as choosing a curve out of the library. Finally, a user may be given a number of answers to a question, such as whether to *"continue"* or *"abort"* an option. The dialogue boxes each contain information about what the user must do. They are rather self-explanatory and easy to use. The menu and dialogue box packages were designed by Trey Matteson.

When an EDGE session begins, the screen will display one large viewing window on the left side and an information window on the right side. The viewing window is where the curves will be drawn; until the user defines a curve, it is empty except for the coordinate axes. The information window contains data about the curve which is currently being displayed. Across the top of the screen are the pulldown menus described above which control the action of the package. At the bottom of the screen is a text output window, which prompts the user for an action, indicates that EDGE is performing calculations, and tells when an error has occurred. In addition to the main viewing window, there are three smaller windows. Because of space limitations, they reside "beneath" the information window. The user has the option of displaying one or the other at any time, and can toggle back and forth between them by picking a button on the screen.

One of the improvements of EDGE over past projects is the use of a *parser*, which allows the user to enter new functions during a session. The parser also allows us to define the position of the coordinate axes in the viewing window. By default, a given window will display the XYZ coordinate system. However, these can be changed to any expression the parser understands. For example, the user might wish to see the curvature plotted over the points of a planar curve, which can be accomplished by setting the third coordinate equal to k. To study the torsion of a space curve one can plot t versus $\tau(t)$.

5.GENERATING OBJECTS

The first thing the user must do is define the curve to be investigated. The user can select from a library of curves in EDGE or type in the equations for a new curve. To examine the helix, the user could choose "helix" from the curve library, or simply type in

$$\{X(t) = \cos(t)\ ,\ Y(t) = \sin(t)\ ,\ Z(t) = t\}\ .$$

Next, the user must indicate a starting and ending value for the variable t , i.e. the range over which the curve is defined. For the helix, a reasonable domain for t would be $[\ -2\pi\ ,\ 2\pi\ ]$, which would yield two complete turns. The user must also indicate the number of increments, i.e. the number of points between the endpoints at which the curve will be evaluated. If the user now tells EDGE to "draw", the helix will appear drawn in the main window.

In addition, EDGE gives the user the option of defining

the curve using "coefficients". For example, the helix can be defined as

$$\{a \cos(t) , a \sin(t) , b t\} .$$

Here the values of the coefficients a and b determine the shape of the helix. The *pitch* of the helix is defined as the ratio
b/a: the greater the pitch, the more stretched out the helix. If
the curve is defined with coefficients, then the user must assign
values before the curve can be drawn. Once the user has provided
this information, the curve will be drawn on the screen, as above.
The machine will conveniently scale the images so that they fit
precisely on the screen (unless an option called "auto-scaling" is
turned off).

   In addition to the curve itself, the user can define "tubes
associated with the curve". The tube is evaluated at each point at
which the curve is evaluated. To define a uniform tube of radius
0,5 around the helix, type

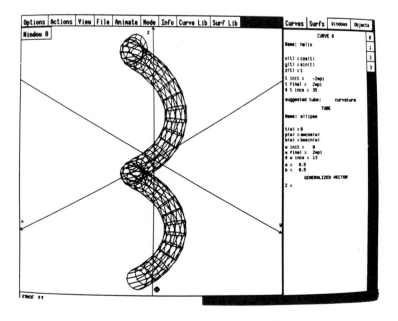

Figure 3

The tube surface of radius 1/2 around a helix.

$$\{0 \ T(t) + (1/2) \ \cos(w) \ P(t) + (1/2) \ \sin(w) \ B(t)\} \ .$$

with $w \in [0 \ , \ 2\pi \ ]$. Here w is an independent variable, like t, with its own domain and its own number of increments. The computer will connect points both around the tube and along the direction of the curve.

Generally, tubes are defined in terms of Frenet frame, T,P,and B. However, sometimes it is useful to define an object dependent on the curve but not in terms of this frame. An example would be to create a second curve by translating by a magnitude a in the direction of the X-axis. The feature, called *generalized vector functions*, allows the user to create this curve by entering

$$X + [a \ , \ 0 \ , \ 0].$$

When a curve is first drawn, it is scaled such that the entire curve will fit on the screen. A feature called *"zooming"* will allow the user to expand a small part of the curve onto the entire screen, so that it is possible to see that smaller portion in greater detail. For example, the spiral generated by

$$\{(1/t) \ \cos(t) \ , \ (1/t) \ \sin(t) \ ,0\} \ , \ .05 \le t \le 20 \ ,$$

is most interesting around the origin. But because X(.05) is so large, what is happening around the origin is barely descernible. Therefore, the user would want to *"zoom"* in on that area by choosing the *"zoom"* option from the menu, drawing a rectangle around part of the curve by depressing the mouse to choose one corner and letting up on the button when the opposite corner has been reached (this process is called *"rubber-banding"*). The area within the box will be expanded to fill the screen. Zooming out is the inverse of zooming in. The user will again create a rectangle. This time the visible portion of the screen will be shrunk into that rectangle, giving a *"zoom out"*.

EDGE has the capacity to view up to four curves at a time. Each curve is defined independent of the others, with its own parameters, coefficients, tube, etc. By default, the screen

will be scaled such that all curves can be viewed in their entire-
ty; this default can be overriden by setting the *auto-rescale*
option to OFF.

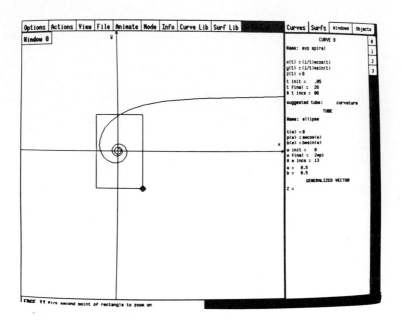

Figure 4
Choosing a rectangle for "zooming in".

Sometimes it is instructive to *"project"* a curve in 3-space
into different planes, especially when a curve is fairly complicated. At these times, the three auxiliary windows are especially
useful. By default, the three-dimensional Cartesian coordinate
system will be displayed in each of these windows. However,
the user can choose to make the window two-dimensional, and
to indicate the actual quantities which are to be graphed
on each axis. Thus, one option would be to project the curve
into the XY , XZ ,and YZ planes, an especially useful way
to study a curve in a neighborhood of the origin.

Figure 5 shows the twisted cubic (t , t^2 , t^3) viewed this way.

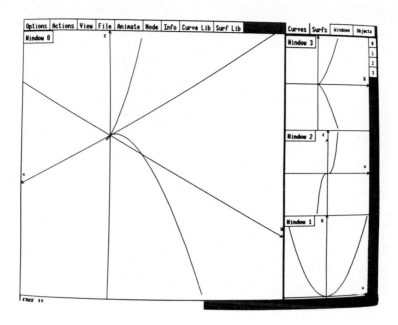

Figure 5

Multiple views of the twisted cubic.

One of the most important features of EDGE is its capacity
to rotate the curve on the screen. When first generated, all
curves are viewed using a previously defined default rotation.
However, while this view may be fine for examining a helix,
it might be inadequate for seeing the *"twisted cubic"* curve. The
user can define a rotation of a certain magnitude about a
coordinate axis or about a screen axis. (The Cartesian axes
will alter as the user rotates. The screen axes remain fixed.
They are defined by letting X give the horizontal screen axis,
Y, the vertical screen axis, and Z, the axis perpendicular
to the screen.) Each time the user enters another rotation,
it is performed relative to the current position. The user
also has the option of going back to the default rotation.

Another particularly useful feature is the animation capability.
Currently, a user can *"animate"* both rotations and coefficient
changhes. In the former case, the user indicates the axis and the
magnitude of rotation as well as the number of *"frame"* for the a-
nimation. The computer will then evaluate the curve once for each
frame, substituting the interpolated value for the magnitude
of rotation. When evaluation is completed, the frames are
drawn on the screen in rapid succession, simulating an animation.

Animating coefficients is an effective way to show how coefficients alter the shape of a curve. The user provides starting
and end values for the coefficient(s) plus the number of frames.
The computer generates the images and flashes them in sequence
on the screen. Animating coefficients is especially useful
in providing information around *"catastrophe points"*.For example,
the curve defined by

$$\{(a + \cos(t)) \cos(t) , (a + \cos(t)) \sin(t) , 0\}$$

has a cusp for a = 1, a loop for 0< a < 1 , and neither for a > 1.
The user can define a five frame animation on the interval [0.5 ,
1.5]. The computer generates images for values of a = 0.5,.75, 1.0,
1.25, and 1.5. When the animation is played back, the user witnes-
ses how the loop appears and disappears as a changes along the
interval, with the cusp appearing in the third frame. The user can
stop at any frame and then move forward or backward one step at a
time.

For any curve, it is instructive to investigate the curvature
and torsion. For example, we might ask where on a curve are the
zeros of the torsion? There are several ways to approach this que-
'stion.

One approach is to create a "torsion tube" around the curve:

$$\{X(t) + 0 \ T(t) + \tau(t) \ \cos(w) \ P(t) + \tau(t) \ \sin(w) \ B(t)\},$$

where w is defined on the interval [0 , 2π] . At each point, a
circle in the plane perpendicular to the curve will be drawn with
radius equal to $\tau(t)$. The zero values can then be observed on the
image.

A curve which illustrates this process is the *space cardioid:*

{(1 + cos(t)) cos(t) , (1 + cos(t)) sin(t) , sin(t)} ,
which has an ordinary cardioid as its projection to the xy-
plane. For this curve, the torsion is zero at exactly two
points. The students can observe this fact on the screen,
and the prove it algebraically.

In Figure 6 we have used one of the most convenient features of
EDGE, the Hidden Edge Removal option, devised by Robert Shapire,
which indicates to the user what lines are "in front of" other
lines. For many views of a space curve, the projection onto a graphics screen will yield crossings or self-intersections even though the curve itself has no double points. Often, it is difficult
to tell which of the parts of the curve or surface are closer
and which are farther away. The Hidden Edge feature provides
that information. At the crossing, EDGE will leave a gap in
the segment which is farther away while leaving the other
segment whole. For the objects of study in our differential
geometry course, the hidden edge procedure is considerably

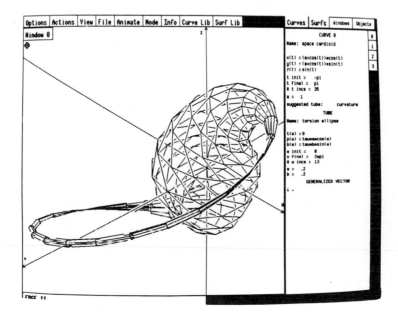

Figure 6
The torsion tube around a space cardioid.

faster than any of the hidden surface algorithms, which would be
too slow to implement interactively in our laboratory.

Another way to study torsion zeros is to use the *"Dynamic Walk"*
or *"flying along the curve"* feature. The idea was suggested by the
work of Michael Freedman on the subject of triply tangent
planes of space curves [6]. We project a curve into its normal
plane at a particular point by displaying:

$$\{-(X(t) - X(t_o)) , (X(t) - X(t_o)) \cdot B(t_o)\}.$$

This projects the entire curve onto the *"windshield"* of a vehicle travelling along the curve with driveshaft in the direction of
the unit tangent $T(t_o)$ and axles pointed along the principal normal $P(t_o)$.

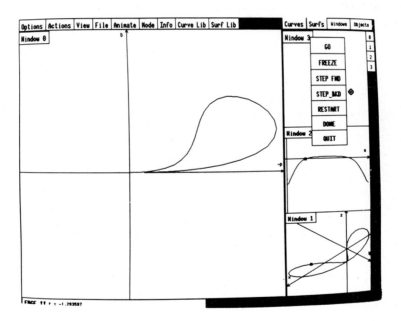

Figure 7

A rhamphoid cusp at a torsion zero of the space cardioid.

The image will always have a singularity at the origin since we
are projecting down the tangent line. If the torsion $\tau(t_o)$ is non-
zero, then the curve passes from the second to the third quadrant

and leave by the same quadrant, obtaining a *rhamphoid* cusp.

The Dynamic Walk process creates one of these projections for each point at which the curve gets evaluated, then displays them in rapid succession (similar to animation). This yields the feeling of "flying along the curve". The user watches the changes from one sort of cusp to another as torsion zeros are encountered.

There is a position for which the projection into the normal plane has three cusps (and therefore there are two other such positions). There is a multiple point of the tangential image curve, in this case a double point which is antipodal to a simple point of the image.

The space cardioid has other interesting properties that are suited to computer graphics investigation. This curve is *self-linked*, in the sense that a small perturbation of the cur-

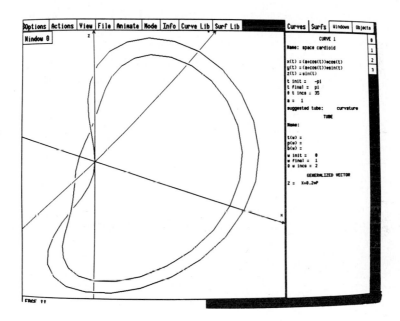

Figure 8

The space cardioid, self-linked with its normal deformation.

all the windows, the amount rotated, translated, and zoomed in all the windows. The user gives the view a name, and it is placed in a library of saved views where it is accessible to that user any time in the future. As with the curve libraries, the saved views can be global or local. The professor and teaching assistant can save views which they would like the class to examine as a common exercise. They also have the ability to take a view from any student and make it accessible to the entire class, providing yet another xay for students to interact with one another's ideas and work. When a *"view"* is retrieved from the library, EDGE makes it possible to continue right where the session left off.

Another feature for transmitting information is the *scratch pad* which allows the user to write thoughts and ideas about the current view. When the view is saved, the scratch pad is saved as well; when the view is retrieved, the scratch pad can be edited anew.

The view-saving feature can be used together with a *homework* system. Using a feature called *class notes,* the professor can assign problems to be examined. The student can respond by generating a few views which illustrate a given issue, including some notes written in the scratch pad. These views can be sent to a speciale homework library accessible to the professor and teaching assistant. They can examine the view, make comments on the scratch pad, and return the commented views to the student. Especially interesting views can be made available to the entire class. These features were used as part of the mid-term examination as well as the final student projects.

7.CONCLUSION

In this report we have discussed many of the features of the interactive program EDGE which we have used in recent years at Brown University. Advances in technology each year bring such interactive programs within the reach of **smaller**

machines and much of what we have described here can be implemented in various ways on inexpensive configurations that are
widely accessible. We are sure that efforts like the interactive
differential geometry laboratory will continue to enhance
the experience of students in all mathematics courses where
geometric intuition can play an important role.

REFERENCES

[1] T.Banchoff, C.Strauss, Real-Time Computer Graphics Techniques
    in Geometry; Proceedings of Symposia in Applied Mathematics,
    Vol.20; The influence of computing on Mathematical Research
    and Education; Amer.Math.Soc. (1974); 105-111.
[2] T.Banchoff, C.Strauss, Real-Time Computer Graphics Analysis of
    Figures in Four-Space; American Association of the Advancement
    of Science Selected Symposium 24 (Westview Press, Colo. 1978);
    159-168.
[3] T.Banchoff, Differential Geometry and Computer Graphics;
    Perspective in Mathematics; Anniversary of Oberwolfach
    (Birkhauser-Verlag, Basel, 1984); 43-60.
[4] T.Banchoff, Computer Animation and the Geometry of Surfaces
    in 3- and 4-Space; Proceedings of the International Congress
    of Mathematicians (Helsinki, 1978); 1005-1013.
[5] T.Banchoff, S.Feiner, D.Salesin, DIAL: A Diagrammatic
    Animation Language; IEEE Computer Graphics and Applications;
    Vol.2, No.7 (1982); 43-54.
[6] M.Freedman, Planes triply tangent to curves with non-vanishing
    torsion; Topology 19 (1980); 1-8.

ECM/87 - Educational Computing in Mathematics
T.F. Banchoff et al. (editors)
© Elsevier Science Publishers B.V. (North-Holland), 1988

GEOMETRIC CRYSTALLOGRAPHY AND COMPUTING IN GEOMETRY

R. FARINATO AND L. LORETO

Dipartimento di Scienze della Terra
Universita' di Roma 'La Sapienza'.

Plane periodic patterns and polyhedra are useful
tools to show how symmetry organizes objects in
space. Further, polyhedra and their related
metrics are geometric entities which associate
concreteness to the mathematical reasoning.
Crystallography provides many situations suitable
for mathematical computing especially when
Interactive Computer Graphics (ICG) are adopted.
The paper illustrates examples from classical and
non-classical Crystallography.

1. INTRODUCTION

Crystallographic research produces material of widely spread
interest in Mathemathics. For a long time 'Mathematical
Crystallography' has denoted that branch of Crystallography
whose content can be approached axiomatically and whose results
are embedded in a pure or applied mathematical environment. The
books [1], [2], [8], [13], [14], [19], [20] and the cross
references therein contained span various pertinent arguments.
The reader is referred to these books for all the
crystallographic background.

Crystallography is geometric by nature so that many parts of
it are suitable also for purposes of educational computing in
Mathematics. From this point of view two subjects, among
others, are of particular interest i.e. the theory of geometric
symmetry and the analysis of polyhedra. We give here a brief
account of principles we are following when applying Interactive
Computer Graphics (ICG) to the production of periodic patterns in
the plane and to the modelling of various kinds of polyhedra in
space.

2. PERIODIC PLANE PATTERNS

Pattern is a term used in extraordinarily different contexts.
As pointed out by Grünbaum and Shepard [10, p.677], [11, p.vii]
a precise and mathematically manageable definition of 'pattern'
was lacking in the literature. We are now indebted to these

Authors for   their considerable work in making 'pattern' a term
usable in a mathematical sense and enriched by meanings not all
strictly dependent only on symmetry. The reader is referred to
[11] for   details and definitions concerning all kinds of
patterns.

   The periodic plane patterns we refer to here are designs P
whose   symmetry   group   is   one   of   17   two-dimensional
crystallographic space groups. Each of these symmetry groups
necessarily contains translations (one of which, at least, is of
minimum length) in more   than one direction in the plane.
Crystallographic space groups   (here called briefly 'plane
groups') belong to the infinite groups. Under all the operations
of symmetry of a plane group no   point of the plane is left
unmoved. Two points of a periodic plane pattern P are called
'equivalent' if one is mapped onto the other by an operation of
symmetry of the symmetry group of P. The   set of all the
equivalent points of P constitutes a transitivity class of P. In
what follows, we shall use simply the words   'plane pattern'
instead of   'plane periodic pattern'. Plane   patterns   are
attractive tools with which to introduce and illustrate, by
computers, concepts   arising from   isometric mapping, group
theory, vector spaces, and so on. There are however many
different ways to use ICG methods to discuss patterns. In our
experience, plane patterns are better understood if some basic
concepts are carefully emphasized at first. The most important of
these are   i) the role of the plane crystal lattice ii) the
action of generators chosen to generate all the operations of the
plane group, iii) the notion of fundamental region and, iv) the
distribution of all the elements of symmetry of the pattern P.

   Focusing on plane crystal lattices is of fundamental
importance, for they constitute the underlying periodic framework
of any   plane pattern P. Other considerations apart, their study
is of   relevant interest in its own right because of the genuine
mathematical nature of the concepts involved. A plane crystal
lattice is the set of the points, called lattice points,
generated by all the integer linear combinations of the two
vectors of a vector basis of the plane. The lattice points thus
form an array of coplanar distinct points always reducible to
have integer coordinates. They are usually represented as a part
of dot pattern. As is well known, only five basically different
(i.e. having different plane group) plane crystal lattices exist,

and it is instructive to be able to explore over and over again
their corresponding dot patterns. A lattice row is any straight
line passing through lattice points. The lattice rows parallel to
the vectors of the basis form a net made up of congruent
parallelograms called 'primitive cells'. Different primitive
cells are possible because the same plane crystal lattice can
be generated by different vector bases. However, all the
primitive cells of the same plane crystal lattice have the same
area. Fig. 1 shows rows and some primitive cells.

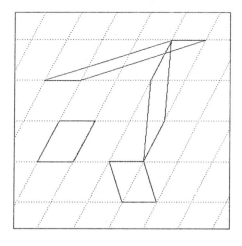

FIGURE 1
Representation of a plane lattice with two
system of rows and some primitive cells
made evident. (All the figs. concerning
patterns are graphic printer output of
what is dispayed, possibly in colour, on
the video screen. The figs. of polyhedra
are plotter output.)

Discussion about the generators of any symmetry group is
highly recommendable. In general, they are not unique, [6].
Although the final result of the production of a plane pattern P
is independent, as a whole, of the choice of the generators of
its plane group, some generators are more convenient than other.
This happens chiefly in many practical instances, especially when
the plane pattern is to be actually drawn, as pointed out by
Schattschneider [18, p.446]. In addition we recall the often
disregarded fact that the same kind of generators used in
different orientations produce different patterns, fig. 2.

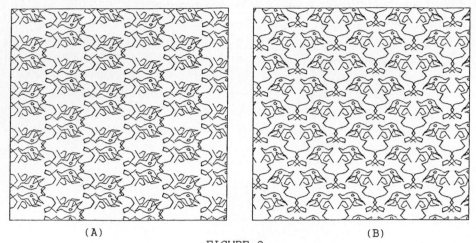

(A)   (B)

FIGURE 2
Effect of the orientation of the symmetry
elements on the same motif. Mirror lines
(space group cm) are horizontal in (A) but
vertical in (B). Note that (B) cannot be
considered as (A) rotated through 90°.

Closely   related to the generators is the problem of the
fundamental regions of a plane pattern, i.e. of the smallest
plane regions within which no point is equivalent to another
point. The   part of a plane pattern P contained in a fundamental
region produces the entire pattern P when the full symmetry of
the plane group of P acts on it. Whatever a pattern may be, its
fundamental regions are   not uniquely defined. Thus, it is good
practice to display some possible fundamental regions in the
drawing. For   example, given the plane pattern of fig. 3 a
possible fundamental region is shown shaded.

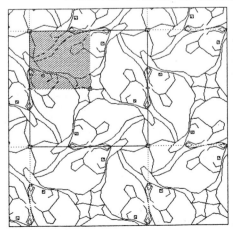

FIGURE 3
Fundamental regions appears
as shaded areas inside the
primitive cell when they
are called by interactive
operations.

When one is looking at a plane pattern, the discovery of its
symmetry elements (translations, points of rotation, lines of
reflection and glide reflection) is often very far from being a
trivial task. So, searching for symmetry elements in periodic
patterns should always be encouraged. However the observer must
be able to check whether the search has succeded. To allow the
match between the elements of symmetry found and those actually
present in a pattern P under examination, it is very useful to
call a symbolic representation, upon P, of its symmetry group.
For example, how to recognize the symmetry elements in the
pattern of fig. 4A may be puzzling but, fig. 4B, the graphical
representation of the plane group used shows how they are
located. Such symmetry interpretations are made much more easy
using colours on polychromatic display screen.[*]

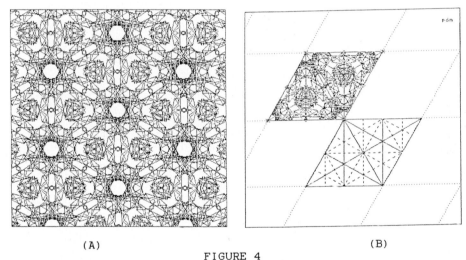

(A)   (B)

FIGURE 4
Symmetry elements of an intricate pattern (A)
become aparent when marked, (B), with solid or
dashed lines (mirror or glide lines) or with
stars, rhombuses etc. (rotation points) In (B)
are shown two primitive cells one with and the
other without the corresponding part of pattern.

The basis of our approach to the ICG production of the plane
periodic patterns is outlined in a preceeding paper [3, p.269].
However, the computer graphics program we now are using is an
updated and completely renewed version of the early one. Further,
the new ICG programs not only allow the choice of one or more
initial motifs but, if desired, a colour can be associated to
each initial motif. This means that every plane group can

---

[*] See page 282 for colour reproductions.

transitively map motifs and their colour. So, one can handle
plane symmetry groups, and the patterns they generate, ignoring
or preserving colours.

3. POLYHEDRA MODELLING

Crystals are formed of matter which becomes polyhedral when
shaped by natural processes. So, polyhedra are three-dimensional
basic geometric objects that crystallographers face everyday.
This fact explains why, in Crystallography, the interest in
polyhedra never falls. Indeed, mathematical research about
polyhedra is far from complete. Even regular polyhedra need more
investigation, [12].

Each polyhedron belongs to a given symmetry group called its
symmetry point group or, briefly, point group. A point group is a
finite group because it does not contain periodic translations
among its elements. Thus, a point group is such that it leaves
unmoved at least one point of the space. In the translational
symmetry groups, treated in the preceeding section, all the
point of the space (plane) are moved. Thus, plane space groups
and polyhedra fit well together to show how symmetry organizes
objects in space.

Polyhedra vary strongly both in shape and in properties. They
range from regular to completely irregular ones, from convex to
non-convex forms and can be simply or not simply connected.

Polyhedra can have constraints in many ways. Regularity is one
of the most remarkable constraints. An interesting and subtle
constraint occurring as a rule in Crystallography is the so
called 'crystallographic restriction', [7, p.60 and 444]. This
means that a 'crystallographic convex polyhedron' cannot have
symmetry axes of order n=5 or n>6 nor can its faces assume any
mutual orientation. It is indeed instructive to demonstrate the
link between the crystallographic restriction on polyhedra and
the periodic plane patterns. Modern Crystallography, however,
accounts today also for icosahedral symmetry groups. Fig. 5 shows
four polyhedra obtained by applying four different point groups
to the same pair of starting planes.

In our experience, we find that some power to pass from one
polyhedron to a slightly or completely different one,
undoubtedly enhances the analysis of the geometric and
topological relationships between polyhedra. So, in the ICG
programs designed for polyhedra modelling we have introduced

many routines   which   allow both the production of polyhedra (of
various classes)   and their interactive modification.

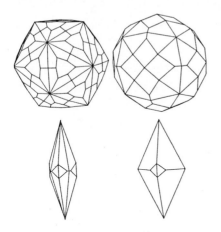

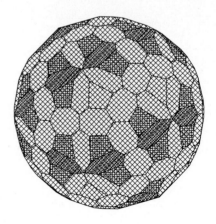

FIGURE 5
Different polyhedra obtained
with different point groups
applied to the same pair of
starting planes.

FIGURE 6
During   the   interactive
description of a polyhedron
some of its faces can be
marked in special mode.

Convex polyhedra are of basic importance. We have considered
two main   methods, [4, sez.A, p. 28.1], to produce the computer
model of a convex polyhedron : 1) starting from the equation of
its faces   or, 2) starting from the coordinates of its vertices.
In most   cases, symmetry considerations and the knowledge of
geometric   constraints greatly simplify computing tasks. Once a
convex   polyhedron is constructed and displayed, then it can be
submitted   to several 'operations'. For instance, one or more
faces can be made more evident, fig.6, or displaced parallel to
themselves (shape modification), fig 7a; vertices can be
truncated (new   faces created) according to their vertex figure,
fig. 7b,; points can be singled out from the vertex figures and
used to produce many other polyhedra. In this way, polyhedra
appear as solid bodies related each to other.

Duality is taken into account too, [16, p.3]. Dual pairs of
convex polyhedra are constructed using the relation of polarity
with respect to a sphere. The polar points of the plane of each
face of a starting polyhedron are considered as vertices of its
dual form: the construction of the corresponding minimum convex
hull is carried out. The dual pairs can be shown as nested

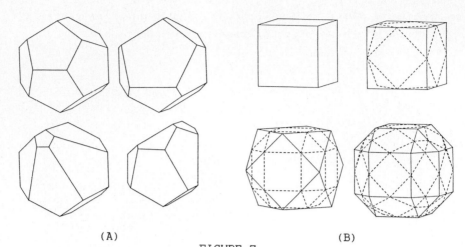

(A)   (B)

FIGURE 7
From one polyhedron to another. (A): how the
displacement of one or more faces may modify a
dodecahedron. (B): recursive application of vertex
figure operations produces new polyhedra.

polyhedra. It is instructive to observe carefully as kinds of
association 'point-plane' other than polarity generate, in
general, wrong results. This happens also in cases of high
symmetry, a situation that seems not always well recognized, [20,
p.72]. Let us consider , for example, the set of centers of the
masses of the vertices of each face of a rhombicdodecahedron (a
Catalan Solid) Q. The corresponding convex hull, [15, p.111], is
a cuboctahedron Q'(an Archimedean Solid), i.e. the right dual of
Q. But the convex hull of the center of the masses of each face
of Q' does not return again Q. The solids in fig. 8 illustrate
this case of false duality. So, experiments with ICG can directly
show which are true dual pairs.

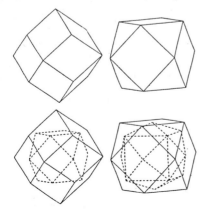

FIGURE 8
True and false duality. Top: the
rhombicdodecahedron,(left), and
its dual, (right). Bottom: the
midpoints of the faces of the
rhombicdodecahedron,(left), give
the true dual whilst, (right),
the midpoints coming out from
the cuboctahedron's faces do not
restore the rhombicdodecahedron
again.

In short, the strategy adopted in all the ICG programs allows computer simulation of a wide variety of convex polyhedra and operative control on them.

A certain number of non-convex solids can be obtained by elevating pyramids on the faces of convex polyhedra or producing each polyhedron's face outside its boundary. By this way, for example, it is possible to derive the first stellation of several convex polyhedra. Disconnected polyhedral forms are easily obtained when only some faces are singled out from a convex polyhedron, fig. 9.

Solid Modelling methods may be profitably used in some instances to explore in depth polyhedral forms and shapes which are not simply connected. This has been done in [9, p. 54].

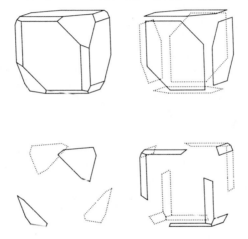

FIGURE 9

Example of disconnected forms obtained by sliding off faces from the original solid shown upper left.

The last, but not the least, remark is concerned with the degree of generality attained in our approach. The polyhedra we handle by ICG methods does not have to be restricted to special cases of particular classes such as, for example, the Platonic or Archimedean or Catalan Solids. Further, the ICG programs are devised to offer the possibility of going beyond mere representation of polyhedra. Metric measurements concerning geometric properties must be made available if one want a complete understanding of polyhedra. To this purpose many menu functions allow access to metrics. For instance, once a convex polyhedron is displayed, one can enquire in real time about dihedral angles between faces, distances between vertices, angles

between edges, perimeter and surface area of each face, total
perimeter, total surface area, total volume, and so on. This
means really comparing polyhedra one with another, [17], [5]. We
have noticed that the computer construction of a few kinds of
polyhedra and, what's more, without any access to geometric
information, favours admiration rather than methematical
reasoning.

AKNOWLEDGMENTS

We wish sincerely to thank Massimo Tonetti and Pino Libonati
who strongly supported the computer production of the figures
concerning patterns and polyhedra respectively. Thanks are due to
all ICGCWG (Interactive Computer Graphics Crystallography Working
Group) members.

NOTES

This paper has been supported by a grant of the Ministero
della Pubblica Istruzione (fondi 40%) and a grant of the
Universita' di Roma 'La Sapienza' (fondi 60% Progetti Ateneo).

REFERENCES

[1] Boisen M. B. and Gibbs G. V., Mathematical Crystallography:
     an introduction to the mathematical fundations of
     Crystallography. P. H. Ribbe (ed.), Mineralogical Society of
     America, vol. 15, 1985.
[2] Brown H., Bulow R., Neubuser J., Wondratschek H., Zassenhaus
     H., Crystallographic Groups of Four Dimensional Space. (John
     Wiley & Sons, New York, 1978).
[3] Cervini L., Farinato R., Loreto L., The Interactive Computer
     Graphics (ICG) Production of the 17 Two-dimensional
     Crystallographic Groups, and Other Related Topics. In: M. C.
     Escher: Art and Science. H. S. M. Coxeter et Al., (eds.).
     Elsevier Science Publishers B. V. (North Holland, Amsterdam
     1986) 269-284.
[4] Ceruti I., Farinato R., Lanza V., Loreto L., Modellazioni 3D
     di Poliedri in Cristallografia. In: Atti del Convegno
     AICOGRAPHICS '85 (Etas Periodici, Milano, 1985) 28.1-28.23.
[5] Cornelli A., Farinato R., Loreto L., Environment of Points
     and Related Polyhedral Configurations: An Interactive
     Computer Graphics (ICG) Approach. Per. Mineral. vol. 53,
     (1984) 135-158.
[6] Coxeter H. S. M. and Moser W. O. J., Generators and relà-
     tions for Discrete Groups. (Springer Verlag, Berlin, 1980).
[7] Coxeter H. S. M., Introduction to Geometry. (John Wiley &
     Sons, New York, 1969).
[8] Engel P., Geometric Crystallography: an axiomatic intro-
     duction to Crystallography. (Reidel Publishing Co.,
     Dordrecht, 1986).
[9] Franchina V., Giorgi E., Loreto L., Solidi Cristallografici
     e loro Combinazioni: un Incontro tra ISM (Interactive Solid
     Modelling) e Cristallografia. In: Atti del Convegno

AICOGRAPHICS   '84 (Gruppo Editoriale Jackson, Milano, 1984)
54–84.

[10] Grünbaum B., Patterns.In: Handbook of Applicable Mathematics
Vol.V part B, Ledermann W, Vajda S.,(eds.), (John Wiley &
Sons.,New York, 1985).

[11] Grünbaum B. and Shepard G. C., Tilings and Patterns.(W. H.
Freeman, New York, 1987).

[12] Grünbaum B. Regular Polytopes– Old and new. Aeq. Mathemati-
cae 16 (1977), 1–20.

[13] Hilton H., Mathematical Crystallography and the theory of
groups of movements.(Dover Publications, New York, 1963).

[14] Jaswon M. A., Mathematical Crystallography, (Longmans, London, 1965).

[15] Lawden G. H., Convexity In: Handbook of Applicable Mathematics Vol.V part A, Ledermann W, Vajda S., (eds.), (John
Wiley & Sons.,New York, 1985).

[16] Libera D. and Loreto L., Platonici, Archimedei, Catala-
ni,..:Costruzione con Metodi di Grafica Interattiva. In:
Atti del Convegno AICOGRAPHICS '84(gruppo Editoriale
Jackson, Milano, 1984), 3–30.

[17] Loreto L., Looking at Crystal Habit and Morphology by Interactive computer Graphics. Per. Mineral. vol.51 (1982) 383–
438.

[18] Schattschneider D., The Plane Symmetry Groups: Their Recognition and Notation, Am. Math. Monthly (1978) 439–450.

[19] Schwarzenberger R.L.E., N–Dimensional Crystallography. (Pit-
man Publishing Limited, London, 1980)

[20] Smith J.V., Geometrical and Structural Crystallography.(John
Wiley & Sons., New York, 1982)

ECM/87 - Educational Computing in Mathematics
T.F. Banchoff et al. (editors)
© Elsevier Science Publishers B.V. (North-Holland), 1988

TEACHING CALCULUS AS A LABORATORY COURSE

Harley Flanders
University of Michigan
Ann Arbor, MI 48109
USA

We shall describe a first course in calculus taught in a
microcomputer laboratory at the University of Michigan.
We shall also describe the MicroCalc software used in
the course. Finally we shall discuss the future of calculus taught as a laboratory science.

1. INTRODUCTION

In the Winter 1987 term, we ran a pilot section of the first calculus course. This was the "mainstream" calculus course; its content is review of analytic geometry and trigonometry, functions
and graphs, limits, continuity, the derivative, applications of
the derivative, the integral, applications of the integral.

The pilot section was taught in a microcomputer lab in which each
student had a workstation, and which contained a projection monitor. This pilot section followed the same curriculum as the 25
other sections of the course, because all students who complete
the first course continue into the second calculus course. Therefore there were limitations on substituting computer work for hand
work and on modifying the course content. The students of all
sections took the same examinations.

2. SOFTWARE

There was no computer programming in the course; the first calculus course in North America is far too intense to allow any time
for a digression into programming. We used an integrated, interactive, calculus software package, MicroCalc 3.0, developed by the
author.

2.1. MicroCalc Content

MicroCalc contains 34 programs, covering almost all topics in the
calculus sequence, plus two utilities and on-line Help. The
programs are interrelated through their sharing common function
data. All of the programs are interactive with just one exception
(Generation of sin x). Each MicroCalc program is menu-driven; the

user chooses options at various points, then enters functions and
numbers.   The programs are written in the structured programming
language Pascal and compiled into machine code.   Many error traps
are built into MicroCalc to insure robustness.   MicroCalc uses
algorithms for graphics, symbolic manipulation, and numerical computation that are standard in the current research literature on
computer science and numerical approximation.

The MicroCalc modules split into four groups and utilities:

Table of Values	Generation of sin x Demo
Graph of y = F(x)	F(x) = 0 by 10-Section
Limits	F(x) = x by Iteration
Difference Quotients	F(x) = 0 by Newton Raphson
Secants and Tangents	Step Functions
Derivatives	Approximate Integrals
Chain Rule	Solids of Revolution
Implicit Differentiation	L'Hospital's Rule
Extrema	Solve F(x) = G(x) Graphically
Taylor Polynomials	Approximate Double Integrals
Polar Coordinates	Approximate Iterated integrals
Parametric Plane Curves	Cylindrical Coordinates
Vector Algebra	Spherical Coordinates
Parametric Space Curves	Vector Calculus
Partial Derivatives	Direction Fields
Surfaces z = F(x, y)	Gradient Fields and Level Lines
Extrema of F(x, y)	Heun Method for ODE's

Help
Scratch Pad
Function Editor

2.2. Hardware Considerations

The microcomputer laboratory contained Zenith 161 computers with
720K core memory, 8087 mathematical coprocessors, and CGA compatible B/W monitors.   Part of the memory was configured as RAMDisk,
containing the complete software code, so no visible disk reads
were noticed when changing from one MicroCalc module to another.
The current release MicroCalc 3.0 works on PC/XT/AT computers.
Various versions support 80x87 coprocessors and CGA, EGA, or HGC
graphics cards.   An Apple IIe/IIc version is also available, and
an Apple Macintosh version is projected for fall, 1988.

2.3 Software Objectives

a) To provide a tool for calculus experiments, thereby providing
      a laboratory environment for the teaching of calculus.

b) To remove the drudgery of calculations, thereby freeing time

for learning the concepts and applications of calculus.
c) To provide a tool for checking the results of hand calculations.

The graphics in MicroCalc provides tools for producing with ease
graphs that could take a great deal of time otherwise. In three
dimensions, MicroCalc will produce curves and surfaces that are
beyond the capabilities of many students and even some instructors.

In preparing **MicroCalc**, the author followed standards which he
believes should guide the preparation of all educational mathematics software:
a) No knowledge of computers should be required to use the software other than the basics of how to insert disks, power on, and
type a few letters.   No computer programming should be required.
b) The documentation should be essentially unnecessary for the
user. What he has to do at any point should be visibly apparent,
and he should never have to stop his work to look up a topic in
the manual. Therefore the manual should be brief, and strictly a
reference work.
c) The software should be interactive, with emphasis on "active."
It should neither spoonfeed nor do "show-and-tell" demonstrations,
but should require user input to produce output, thereby requiring
the user to think about what he is doing, and to be gratified by
the results of his input. He should always feel he is part of the
process.
d) The software should be robust: extremely difficult to crash or
to hang. Input errors should be noted, not punished, with reentry
requested politely.
e) The syntax for input should be as close to the way mathematics
is written as is possible. For instance

    .5 sin 2t   not   0.5*(sin(2*t)).

f) Input of real numbers should allow expressions that evaluate
to reals, so the computer does your calculations for you. For
instance,

    sqrt(1 + sqr .35).

3. COURSE SELECTION AND RESULTS

Knowing that this environment would be frightening to some students, we set the enrollment at 40, higher than usual. At the same time, a parallel section was set smaller, to allow an orderly transfer from the pilot section to the regular section for those students so wishing. The students who came into the pilot section on the first day of class did not know in advance that this section was different from the others. They were told precisely what the situation was and that they could move to the parallel section if they wished at any time (within reason).

About 12 students chose to transfer in the first two weeks, and others left later, so the course ended with 22 students. The performance of the students in the pilot section against the students in all sections of 115 was about the same. Course grades were based on the total of the uniform midterm and final examination scores. For all sections, the average of this total was insignificantly below the average for the pilot section.

4. CONCLUSIONS

First, the computer proved a valuable teaching tool, particularly in situations requiring graphs of functions or where graphs could clarify function behavior.

A second conclusion was that we should separate the lecture/demonstration part of the course from the laboratory part. Our next projected pilot course will consist of three lecture/demonstrations and one supervised laboratory section per week. Part of the homework will be assignments to be done in computer laboratories.

A third conclusion was that we need a laboratory manual to go with the laboratory equipment (the computers and software).

A final conclusion was that some our software modules required modification, and some needed modules were missing. We corrected these deficiencies as we went along.

5. DIDACTIC PRINCIPLES

Traditional material on routine calculation and technique that is easily handled by computers can be pruned from calculus, thereby freeing time to concentrate on the main ideas and theory, and on setting up problems and modelling.

Students at the lower division level learn new concepts and theory most efficiently if they can construct painlessly a rapid succession of examples, and can experiment easily with variations of the parameters in these examples.

Students have much difficulty learning to set up "story problems," that is to build the mathematical model for an applied problem. Current software is no help in learning the skill of building models, but it can compute answers for the student, both sparing him the frustation of lengthy calculations, (as often as not, leading to wrong answers, so he has no reinforcement for his correct set-up), and giving him the time to set up many examples, thereby allocating his learning time to the central issue, not to messy calculations.

Finally, students have great difficulty with three-dimensional situations. They are unable to visualize figures in three space, unable to make the simple drawings of three-dimensional objects that are required in much of calculus, and get little satisfaction from the blackboard efforts of their instructors, who often lack technical drawing skills themselves. Here the computer can be helpful; give it the equation of a surface, and it produces in moments a readable figure that might take a student hours to prepare.

6. FUTURE DIRECTION OF CALCULUS INSTRUCTION

There is currently a great deal of dissatisfaction with the teaching of calculus and with the course content. There are efforts in the United States by professional organizations, private foundations, and government agencies that will surely result in profound changes in calculus.

The computer revolution the been the major cause of this unrest. First of all, because of the computer, discrete and algorithmic mathematics have become much more central in importance, and there are even responsible members of the mathematical community who propose replacing calculus as the major college mathematics core course with a new course centered on discrete mathematics.

Next, calculus is usually regarded by the students taking it as
their hardest course. In spite of generation after generation of
new textbooks, changing the order of the topics, changing the mode
of instruction, and trying everything conceivable, most mathematics departments, and the client departments they serve, are not
satisfied with the results of the calculus teaching. There is a
constant conflict between maintaining standards and lowering student attrition.

We believe that changing calculus into a laboratory course must
come in the next few years, and that well designed use of the
microcomputer in calculus instruction wil be an important factor
in improving calculus teaching, independent of changes in course
content. Central to this use of the computer is software. Indeed, the very existence of software with certain capabilities can
itself influence the curriculum. For example. the possibility of
using a symbolic integration program has caused much discussion on
drastically pruning the long and painful unit on techniques of integration in the second semester calculus course.

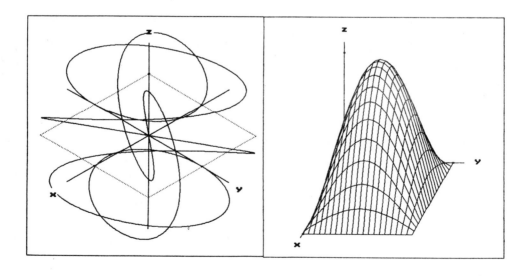

FIGURE 1   FIGURE 2

Examples of MicroCalc graphics

ECM/87 - Educational Computing in Mathematics
T.F. Banchoff et al. (editors)
© Elsevier Science Publishers B.V. (North-Holland), 1988

TEACHING MATHEMATICAL AND PHYSICAL CONCEPTS WITH AN ANALOG SIMULATOR

S.Feliziani
Mathematics and Phisics Department of the University, Camerino (MC) (Italy).

V.Franchina, C.Mengoni, A.Polzonetti
University Inter-Department Computing Center, Camerino (MC), Italy

The implementation of a program that simulates an analog computer on a digital machine has resulted in a simple, yet powerful teaching tool. There is fair evidence that instruction may be greatly improved in some cases when such a method is available. It is well known that several mathematical and physical concepts, especially those involving mechanics, are more easily understood when suitable teaching aids are available. An analog simulator proves to be a superior aid, since the traditional use of digital computers in the educational field still requires a potentially intimidating effort to learn programming languages and operating systems. This is often too complex a road to the final goal; above all there is the risk of focusing more on data processing problems (editing, compiling, loading) than on the goal itself.

1. INTRODUCTION

A program that simulates an analog simulator in a digital computer has recently been implemented [1]. This double-level simulation features such simple and effective interaction that it may be suitable for educational purposes. At this point, it is useful to pinpoint some of the problems that teachers face up when they decide to use instructional aids which rely upon advanced technology.

In this paper we will refer, in particular, to the following scenarios:
- teaching basic concepts regarding mathematical functions and their representation;
- teaching basic concepts in classical mechanics (rigid bodies movement).

As far as mathematics instruction is concerned, use (and abuse) of personal computers is standard in almost any high school (and often even in lower level courses, down to the primary school).

We will not discuss the philosophical and ideological implications of this trend, but it is generally agreed that even the least demanding use of digital computers requires a significant commitment both by teachers and learners. For example, just to plot a function, one has to follow the whole path from writing a source code, editing and compiling it, loading, debugging, down to the output phase. Almost the entire path described is completely uncorrelated to the final goal, which is the function plot: it has no impact on the learner's side. As working with a digital machine is usually a challenging and interesting activity (sometimes frustrating, too), it is easy to conclude that the interest and the efforts of teacher and student alike will go mainly in this direction, and that teaching mathematics will often result in teaching computer science.

This change in focus can be somewhat alleviated through the introduction of special purpose didactic programs, but such programs should still retain a certain degree of interaction, or they will be considered like a videotape. In any case, these special purpose programs must include so many features that non-professional implementation is completely out of the question. This leads to costly, unavailable and obsolescent products, so that in time data processing facilities themselves tend to be forgotten. By using our analog simulator, we think that teachers will have as flexible a tool as a programming language. The chief advantage of the analog simulator is, however, that it is strongly problem oriented. It deals mainly with problems which are described by a set of differential equations, with time as independent variable.

Analog computers would be very well suited for this purpose: unfortunately, they have almost disappeared [2], and their original implementation would have posed so many problems and limitations, including cost, as to make their introduction absolutely unsuitable. This is not true when an analog computer is simulated inside a digital machine, especially when the latter is in the class of personal computers.

2. SYSTEM DESCRIPTION

The already mentioned Ref. [1] gives a detailed description of the software implementation. The following points constitute the highlights of the system:
- all the software, which is available at source level, is written in Pascal and is compatible with UNIX (TM) or MS-DOS environments. The teaching application to which we will refer works with an Olivetti M24 personal machine, with 128 Kb core memory, 360 Kb floppy, and Borland International's Turbo Pascal;
- when started, the analog simulator completely shades out the operating system, so that the user may absolutely forget about the digital environment. The command END will revert operation to the standard operating system;
- only fourteen different commands are needed to introduce or change a wiring schematic, to set up coefficients and constants, to run and stop the simulation, to plot and list results, etc.;
- debugged and tested simulations may be instantly recalled (they reside on disk files), so that it is possible to save current work including parameters set up for future use;
- no operating system knowledge is required;
- expert users may add special operational blocks like function generators,

peripheral devices handlers, and so on to the virtual analog machine.
Because systems which do not accept the Pascal language are still relatively widespread, a FORTRAN77 translation has been attempted, and is currently beta-tested. C language has also been considered as an additional vehicle for dissemination of the simulator.

The well known BASIC language was rejected, however, mainly for reasons of speed. The simulator software is interpretive to some extent, and several table accesses are required at each computing step. These requirements call for both a fast machine (the Olivetti M24 is acceptable, even if not coprocessor equipped) and a highly efficient compiled language.

3. AN APPLICATION EXAMPLE: PLOTTING A FUNCTION

The chosen equation is of the biquadratic type, but any other polynomial form can be solved, as well as a very broad class of more complex functions. The equation is in the form:

$$y = a\ x^4 + b\ x^2 + c$$

The goals of the exercise are to show the shape of the function on the xy plane, to find solutions when y=0, and to inspect shape variation when coefficient values are changed. In particular, absence of relative minimums (when a>0) and real solutions (when y=0) will be shown.

The wiring schematic in Figure 1 uses four integrators, and relies on well established graphic conventions. Initial conditions are also shown in the form of three potentiometers which allow the coefficients to be changed.

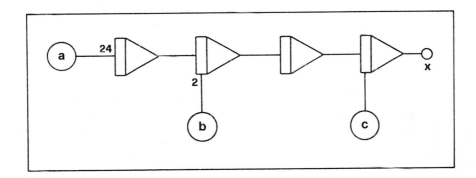

FIGURE 1

Figure 2 shows the command list which led to the above mentioned wiring schematic, while Figure 3 shows several different plots. As the standard operational blocks include direct and inverse trig functions, logs and exponentiation, a widespread class of functions is available.

When functions exhibit discontinuities, it is possible to insert operational blocks which prevent overflows and then to resume normal operation after the critical zone.

```
DEF,IN,Y(3)
DEF,IN,Y(2)
DEF,IN,Y(1)
DEF,IN,Y
DEF,OU,Y_Y
DEF,OU,X_TIME
DEF,PT,A
DEF,PT,B
DEF,PT,C
CON,A,1,Y(3)
CON,Y(3),1,Y(2)
CON,B,4,Y(2)
CON,Y(2),1,Y(1)
CON,Y(1),1,Y
CON,C,4,Y
CON,Y,1,Y_Y
CON,TIME,1,X_TIME
CON,+REF,1,A
CON,+REF,1,B
CON,+REF,1,C
SET,Y(3),1,24
SET,Y(2),4,2
```

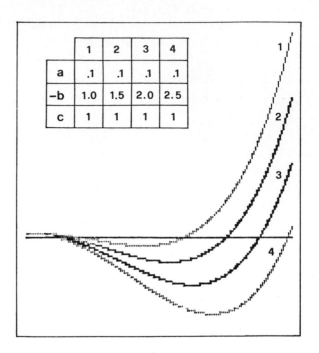

	1	2	3	4
a	.1	.1	.1	.1
−b	1.0	1.5	2.0	2.5
c	1	1	1	1

FIGURE 2   FIGURE 3

Another interesting application consists in plotting several functions which evolve synchronously, with real-time readout. The system is capable of displaying several variables on the same plot, using color to separate curves.

4.   AN APPLICATION TO CLASSICAL MECHANICS: THE IDEAL GUN

The power of the analog simulator is still more evident when dealing with an example from mechanics: the classical ideal gun problem. A shell (in the model, a particle body) is shot by a gun, and the study of its motion in the vertical plane is undertaken.

Let V be the initial velocity, A the shooting angle with respect to the horizontal plane, g the acceleration of gravity, and x and y the coordinates in the trajectory plane. The motion equations are:

$$x'' = 0$$

$$y'' = -g$$

The initial conditions on the first derivatives (velocities) are:

$$x'_o = V \cos(A)$$

$$y'_o = V \sin(A)$$

This leads to the wiring schematic of Figure 4, while Figure 5 lists the commands required for its implementation. Figure 6 shows several "shots", and one can easily find that the maximum distance is covered when the shot is directed at 45 degrees from the horizontal plane (a well known analytical result).

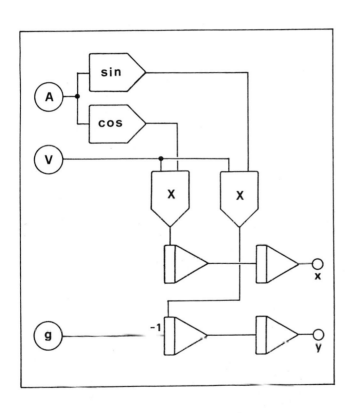

```
DEF,IN,X.
DEF,IN,X
DEF,IN,Y.
DEF,IN,Y
DEF,SN,SINO
DEF,CS,COSO
DEF,MP,VCOSO
DEF,MP,VSINO
DEF,OU,Y_Y
DEF,OU,X_X
DEF,PT,A
DEF,PT,V
DEF,PT,G
CON,A,1,SINO
CON,A,1,COSO
CON,COSO,1,VCOSO
CON,SINO,1,VSINO
CON,V,2,VCOSO
CON,V,2,VSINO
CON,VCOSO,4,X.
CON,VSINO,4,Y.
CON,G,1,Y.
CON,X.,1,X
CON,Y.,1,Y
CON,+REF,1,A
CON,+REF,1,V
CON,+REF,1,G
CON,X,1,X_X
CON,Y,1,Y_Y
SET,X.,1,0
SET,Y.,1,-1
```

FIGURE 4   FIGURE 5

Things may be complicated by adding air resistance in the form of a force proportional to the velocity modulus, and opposed to the velocity vector.

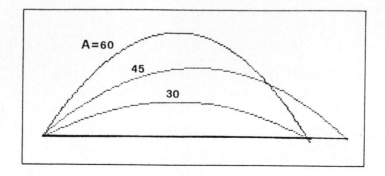

FIGURE 6

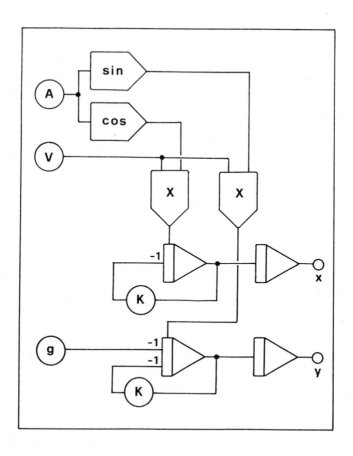

FIGURE 7

Figure 7 shows the wiring schematic for the improved model, which may be described by the following set of equations:

$$x'' = -K\ x'$$
$$y'' = -g\ -K\ y'$$
$$x'o = V\ \cos(A)$$
$$y'o = V\ \sin(A)$$

Figure 8 shows some results, with different values of the initial shooting angle and of the resistance coefficient.

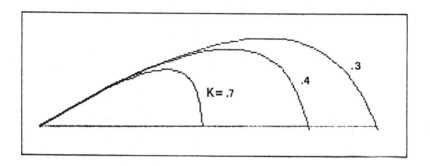

FIGURE 8

5. CONCLUSIONS

The simple examples discussed above should support our feeling that such a simulator may be a powerful, yet simple learning aid. Even non-professional computer users can be immediately proficient when implementing this technique, as no computer language or operating system knowledge is required. A series of tests demonstrating rapid acquisition of the method has been informally performed in universities and high schools.

The opportunity to concentrate on the application, rather than on the technical make-up of the system, was the main advantage which was reported by both users and authors, especially the virtual absence of the debugging phase. This by no means should lead to the conclusion that analog simulation is error-free; however, the extreme synthesis of the mechanization process makes debugging easy and straightforward.

Learners' reaction should be taken into account as well. While it was generally positive, some perplexity was shown by people who had already been exposed to computer knowledge. At first, they felt uncomfortable with a method where the word "algorithm" is practically unknown. This will not necessarily

support the well known opinion of Joe Weizenbaum, who maintains that BASIC
use may impair human brain operation. However...

REFERENCES

[1] Polzonetti, A., Franchina, V., Gagliardi, R., Marcantoni, F.,   Colosimo,
    A., Digital Simulation of Analog Simulators (Proceedings of the EUROPEAN
    CONGRESS ON SIMULATION, Volume B, pag. 105, Prague, September 1987).
[2] Franchina, V., CAD controcorrente: alla riscoperta dei simulatori
    analogici (Proceedings of the ICOGRAPHICS Conference, pag. 37, Milan,
    1986).

ECM/87 - Educational Computing in Mathematics
T.F. Banchoff et al. (editors)
Elsevier Science Publishers B.V. (North-Holland), 1988

MATHEMATICAL EXPERIMENTS ON THE COMPUTER

Ulf GRENANDER

Division of Applied Mathematics
Brown University
Providence, Rhode Island 02912 U.S.A.

ABSTRACT

We shall describe how we have been teaching mathematical software in the Division of Applied Mathematics at Brown University for the last ten years. The main objective is to illustrate the relation between mathematical structures and computer technologies (hardware, languages, input/output devices). The software is intended for mathematical experiments, and we require that the code be flexible and well documented, so that it can be adapted to the varying demands of such experiments.

1. INTRODUCTION

It is notoriously difficult to *involve students in mathematical research at an early stage* of their education. On the graduate level it usually takes a couple of years of preparation before the student can start his own thesis research, led by a supervisor, but to some extent independent. He first has to learn so many techniques, get acquainted with several branches of his discipline, and absorb a large number of abstract ideas and concepts.

This sort of essentially passive learning is necessary, but carries with it some dangers. The student may get an overwhelming impression of mathematics as *scholarship*, but not experience it as an *activity* some elements of which are *guessing*, *heuristic reasoning*, *specialization/generalization*, and so on - topics that seldom make it to the publications.

To compensate for this students are encouraged to do exercises, engage in solving textbook problems. But this is psychologically quite different from real research, working with unsolved problems.

The same is true, even more so, for undergraduate education in mathematics. I would like to describe a teaching experiment that has been going on for more than ten years at Brown University, in which we have tried to involve undergraduates in mathematical research.

We have done this in the format of a senior seminar devoted to mathematical experiments on the computer. *To be able to do such experiments efficiently one needs mathematical software*, but this should only be a tool, indispensable, of course, but not **the main thing.**

The objective should be to facilitate the activities I have mentioned by exploiting current computer technology, hardware, but especially programming languages. At the same time the students should learn some mathematics and apply the knowledge to problem solving in research situations.

This is how I organized this teaching effort together with a colleague, Professor Donald McClure. We select an area of mathematics, teach the students some of the fundamentals of it and discuss how the mathematical concepts and techinques have their computational analogues: data structures, algorithms, and programs.

After such an introduction, necessarily somewhat superficial, we turn to more detailed work and the students are given tasks, the first part of which is to understand the mathematical topic at hand, .and the end product is a carefully tested and documented program that implements the mathematical solution.

2.   To be more concrete, let me mention two examples. The first one, from linear algebra, deals with symmetric, positive definite Toeplitz matrices. Consider the set $\mathcal{R}$ of such matrices and write

$$R = \{r_{s-t};\ s,t=1,2,...n\} \in \mathcal{R}$$

Someone had suggested the following conjecture. If we form the inverse matrix

$$R^{-1} = \{\rho_{s-t};\ s,t=1,2,...n\}$$

it has positive row sums, or, with the vector $e = \mathrm{col}(1,1,...1)$,

$$H : (Re^{-1})_s = \sum_{t=1}^{n} \rho_{s-t} > 0;\ s=1,2,...n.$$

The person who had suggested H, an economist, had collected data, about 25 R-matrices of size 20×20. In all cases H held; 500 positive row sums.

Except for the circulant case, when r_k is a periodic sequence with period n, *little analytical support was available for the conjecture.* In the circulant case, the eigen values and eigen vectors are available in closed form, so that R^{-1} can be done analytically, and the conjecture follows easily. There is also the asymptotic theory of Toeplitz forms which implies that, vaguely speaking, the conjecture holds asymptotically. But the conjecture was not supported in general by any analytical arguments that we could think of. A computer experiment is therefore in order.

Let us write a program that carries out heuristic search for counter examples, generating matrices like R at random, tests for the hypothesis H, and explores the matrix space systematically.

Either we find a counter example, that is a theorem, or the examples may strengthen our intuition by showing what happens at the boundary of the matrix space $\mathcal{R}$.

A technical difficulty is that it is not so easy to pick R-matrices at random. We want them to be symmetric, which is easy, and Toeplitz, also easy. They also have to be positive definite, which is a property further away from our intuition than, for example, symmetry. One necessary and sufficient way of verifying this property (there are lots of others, equally unattractive) is to use *Sylvester's criterion* $\det(R_1) > 0$, $\det(R_2) > 0,...,\det(R_{n-1}) > 0$, $\det(R_n) > 0$ where R_k stands for the k×k section of the R-matrix. Determinants are notoriously clumsy to work with both analytically and computationally, so that instead of generating symmetric Toeplitz matrices "at random" and selecting those that satisfied Sylvester's criterion, *we generated them directly by Carathéodory's representation of finite Toeplitz and positive definite matrices:*

$$r_{st} = \sum_{k=0}^{n-1} f_k \cos \lambda_k(s\text{-}t)$$

where $f_k \geqslant 0$, $0 \leqslant \lambda_k \leqslant \pi$. We generate the fs as i.i.d. variables R(0,1) and the λs as i.i.d. variables R(0,π). The block in the flowchart doing this is called Carathéodory.

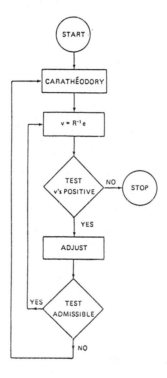

This procedure of generating positive definite matrices directly speeds up the algorithm a lot. We then compute R^{-1} and test whether all entries are nonnegative. If a negative value is found, the program stops. Otherwise it adjusts the λ-vector as in the figure below. The block ADJUST computes the criterion

$$M = \min_s (Re^{-1})_s$$

for two points, say P_1 and P_2 in λ-space. We then move by successive steps in the

direction of decreasing M, along a straight line until we get outside the set of admissible
λ-values, when we start over again generating two new λ-points at random.

Of course, the λ-values outside the region are not really inadmissible (periodicity!),
but we do not want to keep them fixed indefinitely, so this seems a wise choice. A
trajectory in λ-space can then look like P_1,P_2,P_3,P_4. It can also change direction as in
the trajectory Q_1,Q_2,Q_3,Q_4,Q_5,Q_6.

As a modification of this strategy we also varied the f-vector linearly until some
component became negative. We then had a 2n-dimensional phase space.

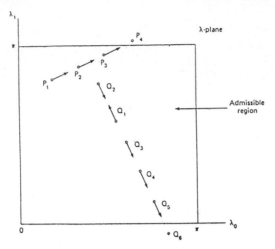

Using the interrupt feature we can stop the program at any time and print out R,
R^{-1}, or M, to give us an idea of what is going on. Notice that *we do not use an entirely
random search* in our 2n-dimensional phase space; that would be very inefficient. *Nor do
we use an entirely systematic search,* which would be nearly impossible, at least in higher
dimensions. Instead a combination of both seemed right.

Executing the program for n = 5 for about 30 iterations produced no negative M.
We noticed a few cases of M-values close to zero, however. For n = 4 about 100 iterations
produced no negative Ms and we did not even find any close to zero. For n = 3,
however, after about 30 iterations the program stopped and produced the matrix

$$R = \begin{bmatrix} 1.6036 & 0.6097 & -0.8127 \\ 0.6097 & 1.6036 & 0.6097 \\ -0.8127 & 0.6097 & 1.6040 \end{bmatrix}$$

with the row vector = (1.8940, -0.8167, 1.8940) and the *conjecture had been disproved*!

To gain some more experience for other values of n, we then returned to n = 4, where
about 300 iterations finally produced a counterexample, and to n = 5, where still more

iterations also gave a counterexample. There was strong evidence that a purely random search would have been wasteful.

Armed with hindsight, it is easy to see what we should have done from the beginning: just evaluate the determinants needed for n = 3 and let the six variables vary, while keeping the matrix nonnegative definite. We finally carried out this boring and time-consuming but perfectly elementary algebraic manipulations. We found, indeed, that there is a nonempty set of matrices R satisfying the conditions and with at least one negative row sum. The set is thin; that is why we did not find it earlier during the experimentation.

Before we had the result of the heuristic search it seemed futile to do this exercise in algebra since the result would probably not have been conclusive, and larger values of n would be too cumbersome. *We had been almost sure that the conjecture was correct, so that our main effort went into unproductive attempts to prove it, not to look for exceptions.*

This experiment also led to some analytical results, but let us only point out that once the strategy of the experiment had been designed (Carathéodory, ADJUST, etc.) it was easily carried out.

3. A more difficult experiment with far-reaching analytical consequences came up in the context of numerical quadrature and integral equations.

In *metric pattern theory* one considers probability densities of the following form, for a vector $x = (x_1, x_2, ... x_n) \in \mathbb{R}^n$,

$$p(x) = \frac{1}{Z} \prod_{i=1}^{n-1} A_\epsilon(x_{i+1} - x_i) \prod_{i-1}^{n} Q(x_i) .$$

Here A_ϵ is an even positive function, Q is a positive function, Z is just a normalizing constant, and ϵ is a small parameter that we shall return to later.

What happens with the probability measure as $n \to \infty$? More precisely, what will be the limiting distribution, if it exists, of x_i, $i = [\alpha n]$, $0 < \alpha < 1$, as $n \to \infty$?

To get it we 'integrate out' all the other x-variables, which implies that we form iterates of the integral kernel K with

$$(Kf)(x) = \int_{-\infty}^{\infty} A_\epsilon(x-y) Q(y) f(y) dy .$$

The kernel is positive, making Frobenius theory a natural tool, and it will come so no surprise that *the limit distribution, say* P_ϵ. *is related to the eigen function* ϕ_ϵ *belonging to the largest eigen value* λ_ϵ *of the kernel, or*

$$\lambda \phi_\epsilon = K \phi_\epsilon$$

Indeed, it had been shown that

$$\frac{P_\epsilon(dx)}{dx} = \text{const.} \times \phi_\epsilon^2(x) .$$

The difficulty comes when we ask how ϕ_ϵ behaves when $\epsilon \downarrow 0$ and

$$A_\epsilon(x) = A\left\lfloor \frac{x}{\epsilon} \right\rfloor$$

with some given function A. To illuminate this question one of the students in the seminar carried out the following computer experiment.

Let us mention in passing, that this and similar limit problems in metric pattern theory are of great practical importance in pattern analysis, for example in image processing.

After learning the basic facts about integral equation, he wrote a program for solving the integral equation $K\phi = \lambda\phi$ numerically. The seminar had developed code for numerical quadrature, but some care was needed in applying it. Just discretizing the equation directly leads to some trouble. Our kernel looks unsymmetric but it can easily be symmetrized by the substitution

$$\sqrt{Q(y)}\ f(y) = g(y)$$

However, the matrix approximating the new kernel is going to be (slightly) unsymmetric anyway leading to trouble.

This and other numerical aspects were taken care of, however, the equation was solved for several choices of the A-function and for decreasing values of ϵ. The results were plotted, a bit crudely, using just a typewriter terminal.

The results were presented in the seminar and generated an enthusiastic response. *The limiting behavior was clearly Gaussian*, which came as a surprise! No central limit theorem seemed to explain it.

This remarkable experimental finding inspired much analytical work that culminated a couple of years later in a complete analytical verification of the conjecture arrived at experimentally. The consequences can still be felt, and it was only this year that a complete understanding of this problem was achieved in terms of the central limit theorem.

4. Our students are asked to write code for various mathematical experiments using the programming language APL. The advantage of APL for this purpose is evident, since its primitives include many of the fundamental structures of mathematics: vectors/arrays, function compositions, sets, operators, etc.

It requires more effort to learn than do the conventional programming languages. Once acquired it enables one to write code in the same spirit as mathematical formulas. It is true that, since the language is usually interpreted, not optimized and compiled, execution can be slow. In our experience this is not a weighty consideration: the user's time should be judged more valuable than that of the machine, now when CPU time has become cheap, and PC's are available on which we can compute overnight if needed.

The disadvantage of APL lies elsewhere: it is not always available on the mainframes of computing centers in universities, and it is not as well known among mathematicians as it deserves to be. Perhaps this will become less important, since many standalone machines, small or medium sized, now have APL as an option.

After students have got some familiarity with APL they usually love it, and enjoy the challenge it poses: it is possible, and often necessary, to be clever writing APL code to an extent quite different from what is the case with the conventional programming languages. It generates enthusiasm among mathematically oriented students and a sense of competition: who can exploit the parallel features of APL in order to achieve the fastest code. This has made the seminar a unique educational experience!

This resulted in some very useful programs and we decided it would be a waste not to save them for our own and other mathematicians' use. We have collected the better ones into what is now an extensive program library. Part of the library has been made public and has been of great use for us.

To give an idea of what sort of program our public library contains I am displaying a table of content:

CONTENTS OF THE APL MATHEMATICAL LIBRARY

I. Utility Programs

    A) Function Editing & Testing (40 UTIL3)

        1) SPLICE - Splices together two APL functions
        2) RENAME - Renames several APL functions at once
        3) TITLE - Changes the name of a function, or adds the lines of fn1 onto the end of fn2
        4) TIMER - Computes amount of CPU time used in executing a program

    B) Formatting Tables (64 UTIL2)

        1) PRINTAB - Formats a set of ordered pairs into a neat, tabular form

    C) Setting up Describe Variables and Editing Functions (66 UTIL1)

        1) COMPOSE - Creates describe variables for APL documentation
        2) EDITDV1 - Edits describe variables
        3) EDIT - Edits APL functions

    D) Displaying Functions and Describe Variables (68 UTIL4)

        1) CMS - Allows user to execute certain CMS commands while in APL
        2) LISTALL - Outputs the code of all APL functions in the active workspace
        3) LISTFNS - Outputs the code of a chosen list of functions
        4) LISTDV - Displays a chosen group of describe variables
        5) SORT1 - Sorts a character matrix alpha-numerically

II. Plotting on TSP/HP Plotters (86 PLOT)

    1) PLOTB - Group of functions for producing graphs
    2) OPS - Group of functions dealing with the linear algebra of piecewise linear arcs

III. Combinatorics (130 COMB)

    1) CONNECT - Finds all connected components of an undirected graph
    2) COMREACH - Computes the commutative reachability matrix of a digraph
    3) CONVERT - Outputs the vector representation of all lines of a digraph

4)   MATMAK - Creates the adjacency matrix of a digraph
5)   REACH - Computes the reachability matrix of a digraph

IV. Descriptive Statistics (131 STATA)

1)   ANALYZE - Calculates mean, variance, mode, median, etc.
2)   FREQTAB1 - Calculates a one-way frequency table
3)   FREQTAB2 - Calculates a two-way frequency table
4)   MODE - Calculates the mode (special formula)
5)   QUANT - Computes sample quantiles
6)   STEMLEAF - Stem and leaf representation of a batch of numbers
7)   TRENDS - Creates two-way frequency table from paired consecutive values of the data
8)   GNNDENS - Group of functions for nearest neighbor density estimation

V. Plotting Histograms and Histosplines (131 HISTO)

1)   HISTOGRAM1 - Prints simple, horizontal histogram at terminal
2)   HISTOGRAM2 - Prints more complicated, vertical histogram at terminal
3)   HISTOSPLINE - Sets up coordinates for the plotting on tsp/hp plotter of an unbounded histospline
4)   BOUNDSPLINE - Sets up coordinates for the plotting on tsp/hp plotter of a bounded histospline
5)   HISTER - Converts a regular histogram to one which can be used by histospline or boundspline
6)   DRAW1 - Draws on the plotter the spline which resulted from histospline or boundspline (also plots corresponding histogram and x-axis)
7)   Several other functions used for plotting

VI. Nonparametric Statistical Testing (133 STATB)

1)   RUNS1 - One-sample runs test for randomness
2)   RUNS2 - Two-sample runs test for randomness
3)   SIGN - Two-sample sign test for comparing prob distributions
4)   KOLMOG - Computes Kolmogorov confidence bands

VII. Fourier Analysis (134 FOURIER)

1)   MRFFT - Mixed-Radix Fast-Fourier-Transform
2)   NINVFFT - Inversion of discrete Fourier transform
3)   PDGRAM - Periodogram of discrete parameter time series
4)   SPECTRI - Spectral est. for discrete parameter time series
5)   TRISMOOTH - Vector smoothing with triangular weights

VIII. Simulation of Discrete Probability Distributions (135 PROBA)

1)   BERN - Bernoulli distribution
2)   BIN - Binomial distribution
3)   DISC - Discrete distribution
4)   DISCRETE - Discrete distribution (alias method)
5)   GEOM - Geometric distribution
6)   MARKOV - Markov chain
7)   MULTINOM - Multinomial distribution
8)   NEGBIN - Negative binomial distribution
9)   POIS - Poisson distribution

IX. Simulation of Continuous Probability Distributions (135 PROBB)

1)   BETA   - Beta distribution
2)   CHI2   - Chi-square distribution
3)   EXPON - Exponential distribution
4)   FISHER - Fisher (F) distribution
5)   GAMMA - Gamma distribution
6)   GAUSS - Gaussian (normal) distribution
7)   PARETO - Pareto distribution
8)   SAMPLE - User defined distribution
9)   SGAUSS - Standard normal distribution

10) STUDENT - Student's t distribution
11) UNIF - Uniform distribution

X. Calculation of Probabilities and Quantiles from Specific Distributions (135) PROBC)

1) BETAR - Incomplete Beta ratio
2) BINFR - Binomial probabilities
3) BINDF - Cumulative binomial probabilities
4) CHISQ - (1-Cumulative Chi-square probabilities)
5) CONVOL - Convolution of two probability vectors
6) FDIST - Cumulative Fisher's F probabilities
7) GAUSSDF - Cumulative Gaussian probabilities
8) NORMDEV - Quantiles of the standard normal distribution
9) POISSONDF - Cumulative Poisson probabilities - for small integers
10) POISSONDF2 - Cumulative Poisson probabilities - unlimited range
11) POISSONFR - Poisson probabilities - for small integers
12) POISSONFR2 - Poisson probabilities - unlimited range
13) TQUANT - Positive quantiles of the Student's T distribution

XI. Demonstration Programs (301 DISPLAY)

1) HEAT - Plots solutions to the heat equation

XII. Prime Numbers (400 PRIME)

1) FACTOR - Prime factorization of an integer
2) NPRIMES - Computes the number of primes less than or equal to a given integer
3) PRIMGEN - Generates all primes between two given integers

XIII. Complex Arithmetic (402 ARITH)

1) ADD, MINUS, MULT, DIVBY
POWER, MAGNITUDE - Arithmetic operations on complex numbers

XIV. Convex Geometry (415 CONVEX)

1) ADDC - Adds polygons K1 and K2
2) ADDC1 - Adds any number of polygons in stacked form
3) AREA - Finds area of a polygon
4) ATAN - Finds angle between two line segments
5) AXF - Computes a rectangle (with sides parallel to axes) to circumscribe any given polygon
6) CIRCUM - Computes a polygon with given face angles which circumscribes a given polygon
7) DCNP - Computes a stacked form array of triangles whose sum i some translation of a given polygon (with no parallel sides)
8) DECOMP - Computes a stacked form array of triangles (and/or line segments) whose sum is a given polygon
9) DEQUAD - Computes a stacked form array of triangles (and/or line segments) whose sum is a given quadrilateral
10) DIST - Computes Hausdorff Distance between K1 and K2
11) DISTANCE - Distance from K1 to K2
12) DISTPT - Maximum distance from a point to a vertex of a polygon
13) DRAWA - Used to plot one polygon on tsp/hp plotter
14) DRAWB - Used to plot more than one polygon at a time on tsp/hp plotter
15) EVALPWL - Evaluates piecewise linear functions
16) GRAPH - Used to produce plots at the terminal
17) HULL1 - Computes the convex hull of a given set of points
18) LINSERIES - For a matrix I, a polygon K and a maximum power n computes

$$K + LK + L^2K + \ldots L^nK$$

19) MATRANS - Computes matrix transformation of a polygon
20) PERIM - Computes perimeter of a polygon
21) **POLY - Converts from support form to standard form**

22) RANDANG - Produces a random angle on the closed interval $(0,2\pi)$
23) REMFVT - Removes false vertices
24) ROTATE - Rotates a polygon by a given angle about the origin
25) SCALE - Scales a polygon by a factor of N
26) SFACE - Converts from raw or standard form to support form
27) STFORM - Converts from raw to standard form
28) SUPPORT - Calculates support function at any given angles in the closed interval $(0,2\pi)$
29) TRANS - Translates a polygon horizontally and/or vertically
30) VADD - Adds elements of a vector
31) WITH - Used in conjunction with the program 'DRAWB' to graph more than one polygon at a time on tsp/hp plotter

XV. Operations on Polynomials (420 POL)

1) ADD, MINUS, MULT, DIVBY POWER, MAGNITUDE - Arithmetic operations on complex numbers
2) MATMULT - Multiplication of two complex matrices
3) GRAMS - Computes coefficients of polynomials orthonormal about a closed curve
4) INTEGRAL - Computes integrals about a closed curve
5) CHOLESKYC - Complex Cholesky decomposition of a Hermitian positive definite matrix
6) INVERSE - Computes inverse of a complex, lower triangular matrix
7) CONFORMAL - Conformal mapping from a plane region to a disc, by Rayleigh-Ritz method
8) POLYC - Multiplies two complex polynomials
9) POLY - Multiplies two real polynomials
10) EVAL - Evaluates several complex polynomials at same point
11) EVAL2 - Evaluates a complex polynomial at several points
12) POLFACT - Factors mod 2 polynomials
13) PDIVBY - Division of one polynomial by another

XVI. Evaluation of Polynomials (421 EVALU)

1) EVAL - Evaluates several complex polynomials at same point
2) EVAL2 - Evaluates a complex polynomial at several points

XVII. Roots of Polynomials (422 ROOTS)

1) SYNDIV1 - Synthetic division
2) SYNDIV - Synthetic division of a complex polynomial by a complex root
3) MULLERM - Finds both real and complex roots of a polynomial equation by Muller's method
4) NEWTON - Uses Newton's method to find a root of an arbitrary function $F(X)$
5) ROOT - Uses a modification of Newton's method to solve for a root of an arbitrary function $F(X)$

XVIII. Computing Orthonormal Functions (423 ORTHO)

1) CHEB1 - Calculates nth Chebyshev polynomial of the first kind at given points
2) CHEB2 - Calculates nth Chebyshev polynomial of the second kind at given points
3) HAAR - Evaluates values of a Haar function at one or several X-values
4) HERMITE - Computes values of orthonormal Hermite polynomial of given order at specified points
5) LAGUERRE - Computes values of orthonormal Laguerre polynomial of given order at given points
6) LEGENDRE - Computes values of orthonormal Legendre polynomial of given order at given set of values

XIX. Calculating Special Functions (423 FUNCTION)

    1)   LOGAM - Evaluates log $(Y(x))$

XX. Systems of Non-linear Equations (424 SYSTEMS)

    1)   DNEWT - Solves a system of n equations in n unknowns

XXI. Numerical Integration (431 INTEGR)

    1)   SIMPSON1 - Approximates a definite integral by Simpson's rule
    2)   TRAPEZOIDAL - Approximates a definite integral by Trapezoidal rule
    3)   INTEGRAL - Computes integrals about a closed curve
    4)   COSFILON - Approximates the integral $F(x) \cos(Tx)$ by Filon's method
    5)   SINFILON - Approximates the integral $F(x) \sin(Tx)$ by Filon's method
    6)   DOUBLE - Approximates double integrals
    7)   GAUSSINT - Approximates integrals by Gaussian quadrature
    8)   GAUSSQ - Calculates the nodes and weights for Gaussian quadrature
    9)   HERMITEQ - Calculates the nodes and weights when the orthogonal functions are the Hermite polynomials
    10)   JACOBIQ - Calculates the nodes and weights when the orthogonal functions are the Jacobi polynomials
    11)   LAGUERREQ - Calculates the nodes and weights when the orthogonal functions are the Laguerre polynomials

XXII. Ordinary Differential Equations (432 DIFF)

    1)   KUTTA - Solves a system of r first order initial value problems numerically using the Runge-Kutta method

XXIII. Interpolation (442 INTERPOL)

    1)   BSPI - Evaluates B-splines at a given point
    2)   SPLDER - Differentiates the cubic spline of SPLINE
    3)   SPLINE - Computes cubic spline coefficients for a given set of points and provides interpolation values where desired
    4)   SPLINT - Integrates the cubic spline of SPLINE

Anyone who would like a detailed description of the library with listing of programs and full documentation can find this in my book "Mathematical Experiments on the Computer" [1].

5. CONCLUSION

We are planning to continue this seminar, since it achieves its purpose to engage undergraduates in mathematical research, and, at the same time, it produces computer code of considerable usefulness for us and other mathematicians that practice experimentation on the computer. At present we are dealing with the area of random geometries for use in image processing. But this requires special hardware, and also differs from the above in that massive CPU time is needed; this will be reported elsewhere.

REFERENCES

[1]   Grenander, U., Mathematical Experiments on the Computer, (Academic Press, New York, 1981).

ECM/87 - Educational Computing in Mathematics
T.F. Banchoff et al. (editors)
© Elsevier Science Publishers B.V. (North-Holland), 1988

DYNAMICAL SYSTEMS

D. Knapp

Department of Applied Mathematics and Centre for Non-linear Studies
University of Leeds
LEEDS LS2 9JT

Abstract
Autocatalytic reactions in chemical or biological systems provide
good examples which motivate and demonstrate the use of numerical
algorithms in topical scientific work.

In recent years I and several colleagues have presented courses to study
the transient and asymptotic behaviour of systems modelled by evolution
equations of the type

$$\frac{dx}{dt} = f(x,\lambda)$$

where $x(t)$ is a variable "*state*" in some finite or infinite dimensional
function space and λ denotes one or more "*parameters*".

Many questions of scientific importance are equivalent to the study of the
asymptotic behaviour of solution paths of dynamical systems as the independent
variable t increases without limit. In particular the "*bifurcation*" values of
parameters λ where asymptotic behaviour changes abruptly are of crucial
interest in science and engineering.

A course in dynamical systems theory introduces the "*invariant sets*" which
may attract solutions asymptotically or act as repellors. These repellors are
still of great significance (although never observed experimentally) since
they act as thresholds or destabilizing influences on the observed solution
paths. Such invariant sets include :

> Equilibrium/steady states
>
> Limit Cycles
>
> Homoclinic orbits
>
> Heteroclinic orbits
>
> "Strange invariant sets"

A course in dynamical systems theory must establish the importance of these
concepts and demonstrate their use in concrete example systems. Numerical
methods and their implementation are natural in this context.

The ideal would be to have an example system which :

1/ *Has been studied experimentally so that it is of real scientific interest.*

2/ *May be modelled by a dynamical system simple enough to be studied analytically initially.*

3/ *Is complex enough to exhibit many different modes of behaviour when parameters are varied and still demonstrate the relevance of analytic investigation.*

The system should also reveal the limitations of analytic study and so motivate and demonstrate the power and utility of numerical methods.

In Leeds one such system has been studied and used in several courses on dynamical systems theory [1].

Two chemicals A, B, react so that the overall result is that the substance A is transformed to a third C, the substance B being unchanged and acting as a catalyst. However the reaction is "autocatalytic" proceeding through the steps:

$$A + B \longrightarrow 3\ B$$

$$B \longrightarrow product\ \ C$$

The reaction proceeds in a stirred tank with a certain flow rate of reactants. The catalyst B may or may not be present in the inflow and may or may not decay in effectiveness as time proceeds (catalyst poisoning).

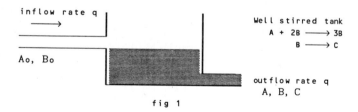

fig 1

We are lead to dimensionless equations:

$$\frac{dx}{dt} = -x\,y^2 + \frac{(1-x)}{\lambda} \qquad (a.1)$$

$$\frac{dy}{dt} = x\,y^2 - \frac{y}{\tau} + \frac{(y_0 - y)}{\lambda} \qquad (a.2)$$

In these equations x, y are dimensionless measures of A, B respectively with the inlet concentration of A used as unit. The inlet concentration of B is y_0, λ is a measure of the flow rate and τ an inverse measure of catalyst decay rate.

In the particular case yo=0 , τ infinitely large (no catalyst poisoning), the steady states at arbitrary values of λ can be completely studied using the analytic solution of quadratic equations. In this case the system behaviour is dependent on the solution of a single differential equation and complete analytic investigation is possible and illuminating.

When $y_0=0$, τ finite, the steady states can again be determined by solution of quadratic equations. This reveals that for some values of τ there is a unique attracting steady state $x = 1$, but there may be a interval (λ_1, λ_2) with multiple solutions due to a branch of solutions not connected to the solution branch $x = 1$ which exists for all flow rates. This branch forms a closed arc in the concentration-parameter space called an "*isola*". Such isolas of steady states (fig 2) are of great interest to Physical Chemists.

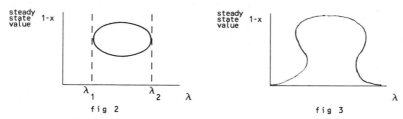

fig 2   fig 3

For general parameter values the steady states are still available as algebraic expressions in the parameters via the analytic solutions of cubic equations but already it is clear that computer assisted graphical representation is more useful. Thus, the use of a microcomputer as an aid to study an analytic expression can be introduced. This can be done with a microcomputer of very modest power.

These graphs reveal that there may now be a connection established between the isola type solution and the other solution to form a "mushroom" type of diagram (fig 3). Thus a diagram of the y_0-τ space can be created indicating those regions where a given type of solution pattern is generated by a selected pair.

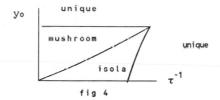

fig 4

Equations (a.1), (a.2) with yo≠0, 1/τ ≠0 also exhibit complex time dependent behaviour away from steady states with steady state multiplicity in some ranges and Hopf bifurcation to stable and unstable limit cycles occurring for some parameter ranges [2]. The topological nature of the solution paths near the steady-state can be determined and leads to the information in fig.5 which shows the Hopf bifurcation point.

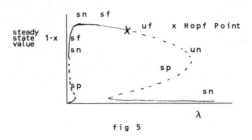

fig 5

In the simple case yo=0 the Hopf bifurcation can be studied analytically and shown to occur when λ takes the critical value

$$\lambda = \left(\frac{1}{2} \right) \tau \left\{ \sqrt{\tau} - 2 + \sqrt{\tau} \sqrt{\sqrt{\tau} - 4} \right\}$$

The stability of the limit cycle produced and its "direction of bifurcation" are also obtainable as analytic expressions but are too complicated to be of any real help. The algebraic manipulation language REDUCE can be used to carry out these analytic procedures but the presence of complicated algebraic processes under the root sign gives expressions less useful than paper and pencil operation. Thus at this point the use of numerical methods is seen to be at least desirable if not totally necessary. These methods [4] reveal that the type of Hopf bifurcation changes at a critical value of (τ ≈ 28) being stable at values in excess of this, unstable at values below. This example can also be used to demonstrate the limitations of symbolic manipulation languages such as REDUCE.

Having shown the analytic structure of a dynamical systems analysis we go on to a system where numerical methods are needed for the initial study of its steady states and their stability. A very useful system for this is the study of nerve impulses in sea squids, with an extensive experimental and theoretical literature.

THE HODGKIN-HUXLEY EQUATIONS

These are the following equations $(1),\ldots,(4)$ for the voltage $V(t)$, Sodium switch-on variable $m(t)$, Potassium switch-on variable $n(t)$, Sodium switch-off variable $h(t)$ when a steady current I is injected into a nerve axon.

$$C \frac{dV}{dt} = I + g_{Na}(V - V_{Na}) + g_K(V - V_K) + g_l(V - V_l) \qquad (1)$$

$$\frac{dm}{dt} = \alpha_m(V)(1-m) + \beta_m(V)m \qquad (2)$$

$$\frac{dn}{dt} = \alpha_m(V)(1-n) + \beta_n(V)n \qquad (3)$$

$$\frac{dh}{dt} = \alpha_h(V)(1-h) + \beta_h(V)h \qquad (4)$$

$$\{\quad g_{Na} = \bar{g}_{Na}m^3h, \quad g_K = \bar{g}_K n^4, \quad g_l = \bar{g}_l \quad \text{each } \bar{g} \text{ constant} \}$$

In these equations $\alpha_m(V)$ etc. are empirical (curve fit) expressions of a complexity which makes analytic study unpromising:

$$\alpha_m(V) = 0.1 \frac{(25 - V)}{\left\{ \exp \frac{(25 - V)}{10} - 1 \right\}} \qquad ,$$

with similar expressions for the remaining quatities β_m, α_n etc.

This example has been used in courses as a system for which it is necessary to use computational methods to make the initial steady state existence and stability analysis. The Newton-Raphson method is used to determine the steady states for a given I value and the QR algorithm used to determine the eigenvalues of the linearisation matrix. After this initial point CONTINUATION ALGORITHMS are introduced to trace out arcs of steady state branches in state-parameter spaces. The code DEPAR due to KUBICEK [4] has been used, and recently we are beginning to use the more powerful code PATH due to Kaas-Peterson [5]. This code will allow continuation of periodic orbits in parameter space and allow the study of period-doubing bifurcations and initial stages of bifurcations leading to "*chaos*".

REFERENCES:

[1] Gray, P., Scott, S. K., Chemical Engineering Science 39, 1087-1097 (1984).

[2] Nonlinear Phenomena and Chaos (Adam Hilger, 1986).

[3] Hassard, B.D., Kazarinoff, N.D., Wan, Y.-H. Theory and Applications of Hopf Bifurcation, (Cambridge University Press: England, 1981).

[4] Kubicek, M, Marek, M. Computational Methods in Bifurcation Theory and Dissipative Structures, (Springer-Verlag, 1983).

[5] Kaas-Peterson, C., Path-Users Guide, (Centre for Nonlinear Studies, Leeds 1987.)

ECM/87 - Educational Computing in Mathematics
T.F. Banchoff et al. (editors)
© Elsevier Science Publishers B.V. (North-Holland), 1988

COMPUTER EXPERIMENTS IN DIFFERENTIAL EQUATIONS

Hüseyin KOÇAK

Lefschetz Center for Dynamical Systems
Division of Applied Mathematics
Brown University
Providence, Rhode Island 02912   U.S.A.*

It is suggested that more geometry and the practice of planned numerical experimentation should be included in both elementary and intermediate-level courses on differential equations. Syllabi for such courses are outlined. A computer program — PHASER: An Animator/Simulator for Dynamical Systems — for implementing these syllabi successfully is described.

1. INTRODUCTION

During the past decade the field of ordinary differential and difference equations has witnessed a remarkable explosion of knowledge, one which has involved both theory and applications to various disciplines from biology to fluid mechanics. Computers have played a crucial role in this development by facilitating detailed analysis of specific systems. Not only have computers suggested new phenomena, as exemplified by the work of Lorenz on strange attractors and the discoveries of Feigenbaum on the universal metric properties of bifurcations of interval maps, but, in a different spirit, they also have enabled Lanford, for example, to complete a mathematical proof of the Feigenbaum conjecture.

It was with the intention of communicating some of this excitement to undergraduate students at Brown University that I initiated, about three years ago, the development of a new course in ordinary differential equations. A major component of this undertaking was to create PHASER: a sophisticated interactive animator/simulator for difference and differential equations for use with IBM personal computers (Figure 1). The first part of this project has culminated in a recently-published mixed-media book-diskette combination [8]. A second book [6], written in collaboration with Professor Jack Hale, is currently in progress and will form the second part. In this paper, I will try to describe both the key features of the computer program PHASER and the ways I used it in instructing first and second courses on ordinary differential equations.

---

*On leave from University of Miami.

PHASER

by

HUSEYIN KOCAK

with the assistance of
L. Heavey, M. Merzbacher
M. Shayegan, M. Sommer
M. Strickman

(c) 1985 by SPRINGER-VERLAG N.Y. Inc.
ALL RIGHTS RESERVED

FIGURE 1. The logo of PHASER.

2. GEOMETRY AND NUMERICS IN A FIRST COURSE

Next to calculus and linear algebra, a first course in ordinary differential equations at the level of, for example, [3] or [4] usually draws the greatest number of students. Since, however, a majority of these students are from disciplines other than mathematics, I believe that the main objective of even such an introductory course should be to present concepts and methods that are useful in analyzing specific equations. An example-oriented approach is equally invaluable for students of "pure" mathematics and future professionals of differential equations. In this regard, one need only cite the impact of the mathematical experiments of Feigenbaum and Lorenz mentioned above (Figures 3, 5 ,7).

To be sure, standard elementary textbooks on differential equations devote a great number of chapters to specific equations. Unfortunately, their presentations frequently consist of collections of tricks and hints for finding explicit solutions, particularly in initial chapters. To this extent such textbooks fail to convey to the student both the spirit and the realities of the subject in its current state. To remedy these shortcomings, I would like to suggest an early injection of a bit more *geometry* and the practice of planned *numerical experimentation* in elementary courses.

On the geometrical side, it is desirable to introduce our subject with a geometrical explanation of what a differential equation is and what it means to solve it. The often-used phrases "a relationship between a function and its derivatives" and "finding a function satisfying the relationship" are rather inadequate. Even in cases where such functions can be found explicitly, mere formulas may reveal limited insight unless they are displayed graphically (Figure 2). Qualitative reasoning and ideas from dynamics should be introduced very early as well. For instance, the concepts of equilibrium points and their stability can easily be discussed on scalar equation. Furthermore, the idea of bifurcation of equilibria is also accessible in this context. The presentation of concepts from dynamics and bifurcations in simple settings can convey to the student a bit more of the modern spirit of our subject.

From the inception of the study of differential equations, it was the realization that most cannot be integrated explicitly which spurred the development of numerical approximation techniques. With the recent proliferation of computers, numerical simulations have become common practice. Because of the inherent difficulty of deciding the accuracy of numerical computations, however, some "purist" educators may discard numerics as inferior mathematics. This need not be so. Numerical solutions of differential equations can indeed be a source of "good" mathematics. After all, numerical integration of a differential equation is simply the iteration of an appropriate map. For example, consider the scalar logistic differential equation describing the growth of a single population:

$$\dot{x} = ax(1 - x),$$

where a is the intrinsic growth rate of the population. Using Euler's algorithm with step size h, after scaling appropriately, the logistic differential equation becomes the logistic map

$$x_{n+1} = \lambda x_n(1 - x_n),$$

where $\lambda = 1 + ha$. Now, investigation of the effect of the step size on the accuracy of numerical solutions leads naturally into discussion of the marvelous dynamics of this map, which are particularly amenable to study on the computer (Figure 3). As it is evident from this simple example, I found it illuminating to include a short discussion of difference equations, or iteration of maps, in conjunction with a presentation of numerical methods, even in a first course in differential equations.

As I have indicated above, most elementary textbooks on differential equations, at the moment, are rather close to each other in spirit that is not entirely satisfactory from the viewpoint of geometry and numerical experiments. My major motivation for developing

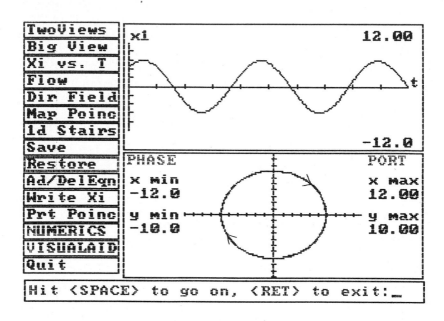

FIGURE 2ab. *Geometry of the linear harmonic oscillator:* (a) Text of equations and current setup, (b) Component of a solution versus time, and its orbit on the phase plane. (*Continued*)

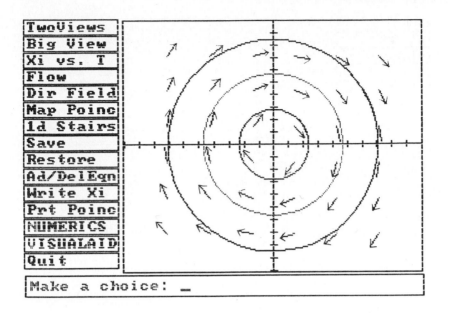

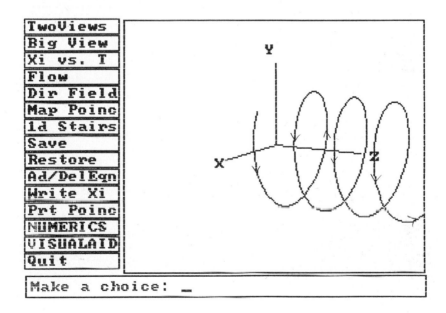

FIGURE 2cd. *Harmonic oscillator continued:* (c) Direction field and the phase portrait, (b) A trajectory in (t, x_1, x_2)-space.

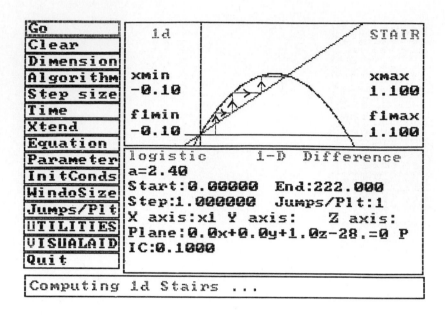

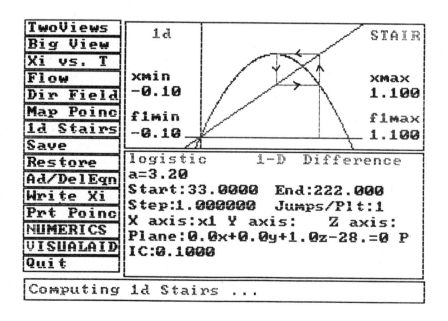

FIGURE 3ab. *Period doubling in the logistic map:* (a) Attracting fixed point,
(b) Attracting period-two orbit (transients discarded). (*Continued*)

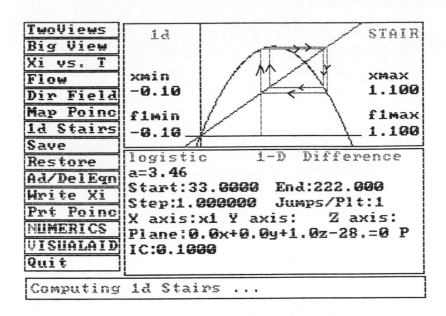

```
TwoViews       1d                          STAIR
Big View
Xi vs. T
Flow        xmin                           xmax
Dir Field   -0.10                          1.100
Map Poinc   f1min                          f1max
1d Stairs   -0.10                          1.100
Save
Restore     logistic      1-D  Difference
Ad/DelEqn   a=3.46
Write Xi    Start:33.0000   End:222.000
Prt Poinc   Step:1.000000   Jumps/Plt:1
NUMERICS    X axis:x1 Y axis:    Z axis:
VISUALAID   Plane:0.0x+0.0y+1.0z-28.=0  P
Quit        IC:0.1000

Computing 1d Stairs ...
```

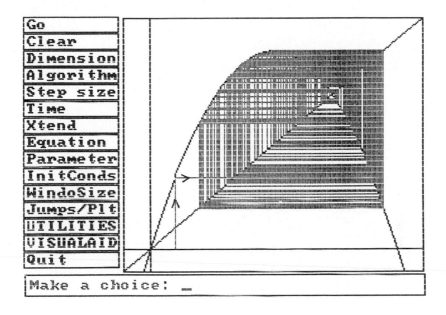

```
Go
Clear
Dimension
Algorithm
Step size
Time
Xtend
Equation
Parameter
InitConds
WindoSize
Jumps/Plt
UTILITIES
VISUALAID
Quit

Make a choice: ▄
```

FIGURE 3cd. *Logistic map continued:* (c) Attractive period-four orbit, (d) Non-periodic orbit — Chaos.

PHASER was, in fact, to remedy this situation by providing some of the necessary tools for geometric study and for numerical simulations of difference and ordinary differential.

3. PHASER

PHASER is an interactive graphics-oriented program designed to help create, manipulate, and store various features of ordinary differential and difference equations. It is written for IBM personal computers and accompanies the book [8]. Despite its sophistication and versatility, PHASER requires no programming knowledge and therefore quite easy to use.

With the help of a menu, the user first creates a suitable window configuration for displaying a combination of views — phase portraits, stair step diagrams for scalar maps, texts of equations, Poincaré sections, etc. For instance, the upper portion of the screen can be used to display a solution in the phase plane, including the vector field, while the lower portion contains the plots of variables versus time. Next, in preparation for numerical computations, the user can specify various choices from another menu. He or she can choose, for instance, to study from a library of over sixty equations ranging from the most rudimentary to ones suitable for research. Of course, one of the selections on the menu enables the user to enter new equations into the library without programming. Once an equation is chosen for investigation, its solutions can be computed with a variety of numerical algorithms, step sizes, etc. Moreover, parameters in the equation can be changed interactively to explore possible bifurcations.

In addition to numerics, PHASER provides the user with extensive graphical tools. Images can be projected from higher dimensions into three and then manipulated graphically in various ways. For instance, they can be rotated, zoomed in, and viewed with perspective projection, or they can be sliced with a desired plane to produce Poincaré maps. These graphical aids are invaluable in sharpening the geometric intuition of the student.

During simulations, solutions can be saved in many ways: as a hard copy of the screen image, as a printed list, or in a form that can be reloaded into PHASER at a later time. An added feature is the animation facility for preparing classroom demonstrations or exercise sets for students. As each image is created, all the data necessary to recreate it can be stored in a disk file. Data for successive images can be appended to this file. Later, by retrieving the file, the screen images can be reconstructed as a dynamic animated sequence.

4. AN OUTLINE FOR A SECOND COURSE

Both the main suggestions I have advocated above — geometry, numerical experimentation, dynamics and bifurcations — and the potential of PHASER can, of course, be realized more fully in a second course in ordinary differential equations. A definitive syllabus for such a course, however, is yet to emerge. Several existing textbooks [1, 2, 7] on differential equations and [5] on maps offer attractive choices. As a result of my experiences in teaching a second course, in collaboration with Professor Jack Hale, to Brown undergraduates several times, I would like to offer here a somewhat different alternative. The following syllabus is essentially the table of contents of our textbook [6], that is currently in progress:

- *Scalar autonomous equations*: Geometry of flows, stability of equilibrium points, examples of elementary bifurcations, numerical computation of bifurcation diagrams, equivalence of flows. Scalar maps and Euler's algorithm, bifurcation of fixed points, the Logistic map.

- *Scalar nonautonomous equations*: Geometry of periodic equations, periodic equations on a cylinder, stability and bifurcations of periodic solutions, Poincaré maps, averaging, equations on a torus, circle maps.

- *Planar autonomous equations*: Geometry of flows, examples of elementary bifurcations, linear systems, reduction to canonical forms, topological equivalence and bifurcations in linear systems, stability and instability of equilibrium points from linearization, Liapunov functions, preservation of the saddle – stable and unstable manifolds.

  Center manifold theorem, saddle-node bifurcation, Poincaré-Andronov-Hopf bifurcation, computation of bifurcation curves, Poincaré-Bendixson theorem, stability and bifurcations of periodic orbits, conservative and gradient systems and their perturbations, homoclinic orbits and their bifurcations.

  Planar maps: linear maps, simple numerics, linearization, saddle-node and period doubling bifurcations, Hopf bifurcation, examples of complicated behavior.

- *Equations in three and four dimensions*: Examples of bifurcations of equilibrium points and periodic orbits, forced oscillators, strange attractors in three dimensions, Hamiltonian systems in four dimensions, pair of harmonic oscillators, examples of integrable and nonintegrable systems.

Because of the inherent difficulty of the subject matter, our approach is to introduce the ideas from qualitative dynamics and bifurcations listed above in as simple a setting as possible. We state the definitions and theorems precisely, but omit some of the difficult or unprofitable proofs. We do make a determined effort, however, to illustrate mathematical concepts and facts with concrete equations. We also regularly supplement the lectures with computer experiments using PHASER both to develop the geometric intuitions of students and to facilitate the analysis of the particularly complicated equations.

5. PHASER IN INSTRUCTION

Modesty aside, I am happy to be able to report that PHASER has been well received by students and colleagues who have used it in both first and second courses on differential equations. PHASER has been reported to be an invaluable aid in realizing the main goals I have advocated above: geometry, numerics, dynamics and bifurcations in specific systems.

For instructional purposes, PHASER has been used in two logistically different ways. The first has been to supplement lectures with selected computer experiments illustrating the dynamical phenomenon under study. For this purpose, it is sufficient to equip a classroom with an IBM personal computer connected to a projector. This setup is rather inexpensive, as a good monochrome projector costs less than one thousand dollars. The second way requires the active participation of students in a computing laboratory equipped with half a dozen personal computers. Here, students carry out their weekly assignments and individual projects under the direction of an assistant. It has been comforting to observe that practically no student experienced initial fear of the computer and that within the first hour all were ready to explore the program independently. The results of these two complementary uses of PHASER have been most satisfying, particularly as many students later return to the laboratory on their own initiative.

6. THE TWO FACES OF EXPERIMENTAL DYNAMICS

The subject of dynamical systems is inherently difficult and the danger of oversimplification is quite real. It is easy for students to get the false impression that computers can provide answers to all their questions. To guard against this, situations where numerical simulations are misleading should be pointed out (Figure 6). It should also be impressed upon students that in somewhat complicated equations, not much can be gained by simply plotting orbits at random. To succeed, there must be a reasonable understanding of the mathematical phenomenon one hopes to observe, and one must be willing to compute other dynamical quantities, such as invariant manifolds, Liapunov exponents, etc.

On the positive side, some simple experiments can spark the interest of students and instructors alike by bringing out the dynamics in dynamical systems. For example, one can first display two solutions of the oscillator of van der Pol simultaneously, and observe their

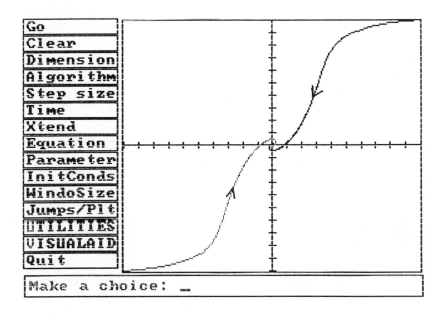

```
TwoViews │           vanderpol
Big View │   Oscillator of van der Pol
Xi vs. T │ x1' = x2 - (x1^3 - a * x1)
Flow     │
Dir Field│ x2' = - x1
Map Poinc│
id Stairs│
Save     │
Restore  │ vanderpol    2-D  Runge-Kutta
Ad/DelEqn│ a=1.00
Write Xi │ Start:0.00000   End:111.000
Prt Poinc│ Step:0.030000   Jumps/Plt:1
NUMERICS │ X axis:x1 Y axis:x2 Z axis:
VISUALAID│ Plane:0.0x+1.0y+0.0z+0.0=0 P
Quit     │ IC:0.2000 0.2000

Make a choice: __
```

```
Go       │
Clear    │
Dimension│
Algorithm│
Step size│
Time     │
Xtend    │
Equation │
Parameter│
InitConds│
WindoSize│
Jumps/Plt│
UTILITIES│
VISUALAID│
Quit     │

Make a choice: __
```

FIGURE 4ab. *Hopf bifurcation in van der Pol's oscillator:* (a) Text of equations and current setup, (b) Origin is hyperbolic and attracting. (*Continued*)

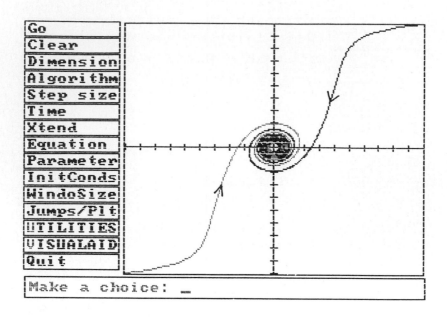

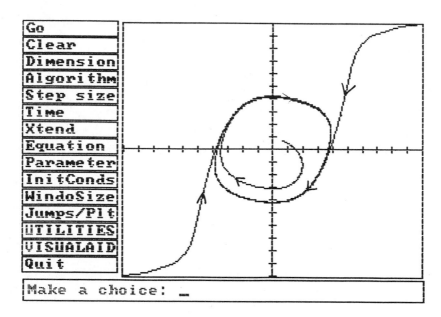

FIGURE 4cd. *van der Pol continued:* (c) Origin is nonhyperbolic but still attracting, (d) Birth of an attracting limit cycle.

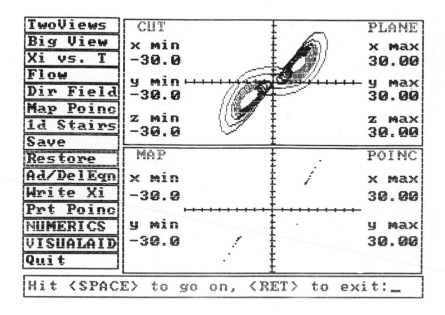

```
TwoViews    |           lorenz
Big View    |     The Lorenz equations
Xi vs. T    | x1' = s * (-x1 + x2)
Flow        |
Dir Field   | x2' = r* x1 - x2 - (x1 * x3)
Map Poinc   | x3' = -b * x3 + (x1 * x2)
1d Stairs   |
Save        |
Restore     | lorenz        3-D  Runge-Kutta
Ad/DelEqn   | s=10.0 r=28.0 b=2.66
Write Xi    | Start:0.00000  End:16.0000
Prt Poinc   | Step:0.030000  Jumps/Plt:1
NUMERICS    | X axis:x1 Y axis:x2 Z axis:x3
VISUALAID   | Plane:0.0x+0.0y+1.0z-28.=0 P
Quit        | IC:5.0000 5.0000 5.0000

Hit <SPACE> to go on, <RET> to exit:
```

```
TwoViews    | CUT            |          PLANE
Big View    | x min          |          x max
Xi vs. T    | -30.0          |          30.00
Flow        |                |
Dir Field   | y min          |          y max
Map Poinc   | -30.0          |          30.00
1d Stairs   |                |
Save        | z min          |          z max
Restore     | -30.0          |          30.00
Ad/DelEqn   | MAP            |          POINC
Write Xi    | x min          |          x max
Prt Poinc   | -30.0          |          30.00
NUMERICS    |                |
VISUALAID   | y min          |          y max
Quit        | -30.0          |          30.00

Hit <SPACE> to go on, <RET> to exit:
```

FIGURE 5ab. *Strange attractor of Lorenz:* (a) Text of equations and current setup, (b) A nonperiodic orbit and its Poincaré map.

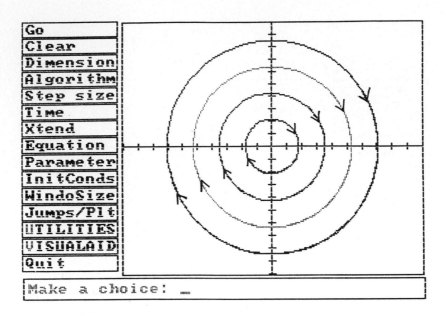

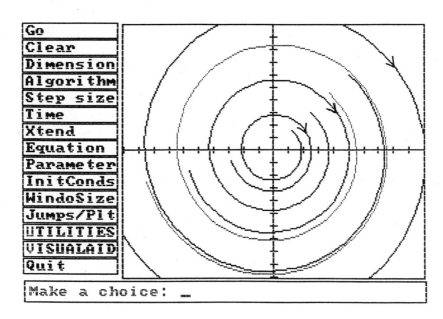

FIGURE 6ab. *Ill effects of numerics on the computation of the phase portrait of the linear harmonic oscillator:* (a) With Euler, (b) With Runge-Kutta.

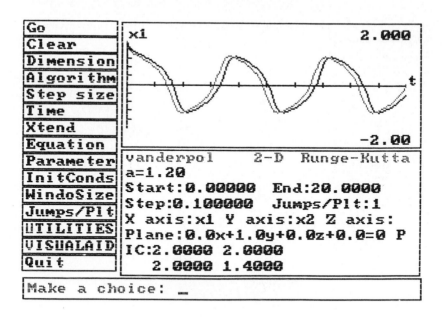

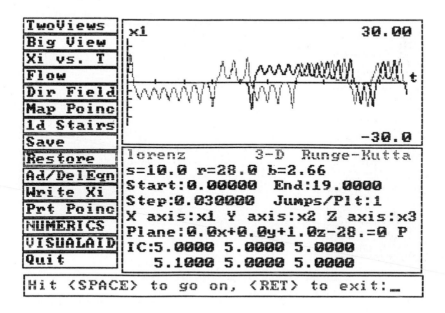

FIGURE 7ab. *Dependence on initial conditions:* (a) Insensitivity in van der Pol, (b) Sensitivity in Lorenz.

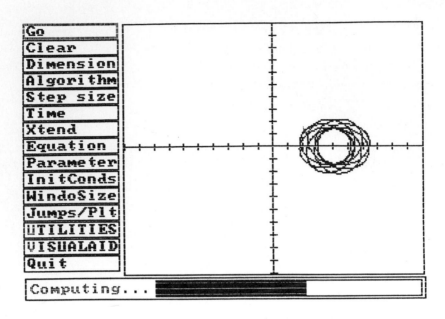

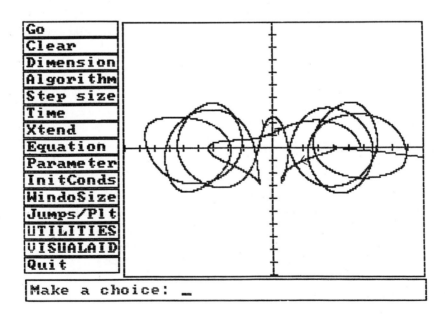

FIGURE 8ab. *Two orbits of the restricted three-body problem:* (a) Capture by a primary, (b) Shuttling between the two primaries.

asymptotic phase on the unique limit cycle. If one then runs two solutions of the Lorenz equations starting close to each other, again simultaneously, it becomes readily apparent how quickly they do rather different things (Figures 4, 5, 7). Although one can prove the observation for van der Pol, it is quite difficult to do any better for Lorenz. My experience, gained through repeated presentations of this sort, shows that when used carefully, interactive computer graphics is a valuable tool for performing numerical experiments and illustrating theoretical precepts with concrete examples.

Carefully planned numerical experiments coupled with geometrical insight, I believe, is a superior way to prepare students to apply techniques from dynamical systems to real-life problems (Figure 8). Moreover, it is indisputable that computers will play an increasingly prominent role in pure and applied mathematics; dynamical systems is a 'natural' place for learning how to use them more effectively. *

ACKNOWLEDGMENTS

I would like to thank many students and colleagues at Brown and Miami for their invaluable help in the development of PHASER and for their enthusiastic participation in courses. It is with particular gratitude that I wish to acknowledge the guidance and friendship of Professor Jack Hale who profoundly influenced both PHASER and my outlook on mathematics as a whole. Finally, I would like to thank Professor M. Emmer for his kind invitation to ECM/87, Rome, which prompted the writing of this paper.

REFERENCES

[1] Arnold, V., *Ordinary Differential Equations* (The MIT Press, Cambridge, 1980).
[2] Arrowsmith, D.K. and Place, C.M., *Ordinary Differential Equations* (Chapman and Hall, London, New York, 1982).
[3] Boyce, W.E. and DiPrima, R.C., *Elementary Differential Equations and Boundary Value Problems, Fourth Edition* (John Wiley & Sons, New York, 1986).
[4] Braun, M., *Differential Equations and Their Applications, Third Edition* (Springer-Verlag, New York, 1983).
[5] Devaney, R.L., *An Introduction to Chaotic Dynamical Systems* (Benjamin-Cummings, Menlo Park, 1986).
[6] Hale, J. and Koçak, H., *Differential Equations: An Introduction to Dynamics and Bifurcations*, in preparation.
[7] Hirsh, M.W. and Smale, S., *Differential Equations, Dynamical Systems, and Linear Algebra* (Academic Press, New York, London, 1974).
[8] Koçak, H., *Differential and Difference Equations through Computer Experiments*, with diskettes containing *PHASER: An Animator/Simulator for Dynamical Systems* (Springer-Verlag, New York, 1986).
[9] Koçak, H., Merzbacher, M., and Strickman, M., "Dynamical Systems with Computer Experiments at the Brown University Instructional Computing Laboratory" (Technical Report No. CS-84-14, Department of Computer Science, Brown University, 1984).

---

* See page 283 for colour reproductions.

ECM/87 - Educational Computing in Mathematics
T.F. Banchoff et al. (editors)
© Elsevier Science Publishers B.V. (North-Holland), 1988

USING COMPUTERS IN CALCULUS EXAMPLES-CLASSES FOR ENGINEERS

Maria MASCARELLO, Anna Rosa SCARAFIOTTI

Dipartimento di Matematica, Politecnico di Torino
Corso Duca degli Abruzzi, 24
10129 Torino, Italy

We report on a didactical research project at the Polytechnic
of Turin, concerning the use of the computers as an aid for
the first two years of the basic mathematics courses.
As an example, we discuss the contents of an exercise-session
in Mathematical Analysis II, on complete elliptic integrals
of the first and second kinds.

1.   At the Polytechnic of Turin an introductory computer laboratory (LAIB) is
at   the   disposition   of the students;   for the first   two   years   Engineering
students   have   access to the laboratory after the completion   of   a   specific
course   which   prepares   them for a meaningful utilization   of   the   available
calculating devices and also supplies them with a good knowledge of a language
(BASIC, PASCAL, FORTRAN) ([1], [6] and [7]).

Several   professors   of the Faculty of Engineering have been   experimenting
the   use   of LAIB as an aid for the first two years of the   basic   mathematics
courses: from the resulting experience two didactic strategies have emerged.
These were asking the students themselves for elaborate the software,   or else
using   existing   software   packages on the market.   It was observed   that   the
preparation of software is,   even from a mathematical standpoint,   an occasion
for investigation of the topic at hand. However, a certain risk was noted in a
tendency   of   interest towards the computer itself to the   detriment   of   time
intended for dedication to mathematical reflection. As far as concerns the use
of   already available software packages,   the possibilities are   many. We have
readily   available   software   elaborated by collegues   teaching   in   analogous
courses   in   other faculties both in Italy and abroad,   that were produced   by
students of previous courses and, of course, software offered by the companies
producing calculating devices.

The following is the presentation of an exercise developed in the course of
Mathematical Analysis II on the calculus of Elliptic Integrals; the didactical
strategy   followed   is the first,   the elaboration of the   software   entrusted
completely to the students.

2.   For a long time the study. or at least the definition, of the complete
elliptic integrals of the first and second kins has been part of the programme
of Mathematical Analysis II in the Faculty of Engineering at the   Polytechnic
of Turin ([2]). It is given, particularly, in the classical form

$$(1) \qquad K(k) = \int_0^{\pi/2} (1-k^2\sin^2 x)^{-1/2} dx \ , \quad 0 < k^2 < 1$$

the complete elliptic integral of the first kind, and

$$(2) \qquad E(k) = \int_0^{\pi/2} (1-k^2\sin^2 x)^{1/2} dx \ , \quad 0 < k^2 < 1$$

the complete elliptic integral of the second kind.
    The study of these functions occurs in the chapter concerning computation
of definite integrals. As is known, one may arrive at (2) wanting to calculate
the length L of the ellipse of the semiaxis a and b , with

$$(3) \qquad L = 4aE(k) \quad \text{where} \quad k = (a^2-b^2)^{1/2}/a$$
$$\text{is the eccentricity of the ellipse.}$$

    Here, L is approximated by the circumference of a circle with radius

$$(4) \qquad r = 3(a+b)/4 - (ab)^{1/2}/2$$

with an error δ satisfying

$$(5) \qquad |\delta| \leq 0.4 \ k^8/(1-k^2)$$

([8], pag. 332).
    The following is the first part of the exercise at the microcomputer   in
which the student is asked to evaluate   L   numerically, using different
procedures and comparing the final results at the end:
a) Consider a polygon inserted in the ellipse and increase progressively the
number of the sides of the polygon. The length of the perimeter of the polygon

gives an approximation of L . This procedure is visualised on the screen.

b) The integral E(k) is evaluated by means of a formula of quadrature, for example the formula of Boole, increasing the number of subdivisions of the interval of integration [0,π/2].

c) The integral E(k) is evaluated using the procedure of integration by series, increasing the number of the terms in the partial sum.

d) One approximates L by the circumference of a circle of radius r as in (4). The initial ellipse and the circle above are visualised also on the screen (see Figure 1).

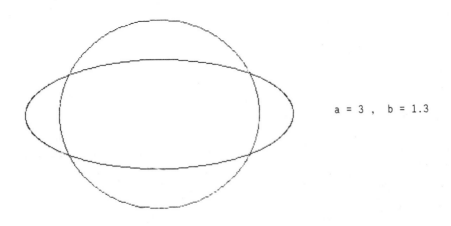

a = 3 , b = 1.3

FIGURE 1

Comparing the results, the methods b) and c) seem best.

As a first application of the integral (1), let us remember how K(k) appears in the study of motion (without friction) of a simple pendulum whose oscillations are not considered infinitely small. The motion of the pendulum results in a periodic motion with perid T given by

(6) $\quad T = 2(1/g)^{1/2}K(k)$

where k = sin α , 0 ≤ α ≤ π/2 , 2α being the maximun elongation of the pendulum , l its length and g the acceleration of gravity ([9]).

In the second part of the exercise the student is invited to evaluate K(k). For this calculation procedures applied are definitely a) and b), already used for the integral E(k) . However, we obtain much better results, by using the algorithm of Gauss for the arithmetic-geometric mean agM, giving the formula

(7) $K(k) = \pi/(2M(1,(1-k^2)))$.

To define agM we begin with the inequality $(a+b)/2 \geq (ab)^{1/2}$
between arithmetic and geometric means. Notice that from one pair of positive
numbers,   a and b , we obtain a second pair, $(a+b)/2$ and $(ab)^{1/2}$. If we
iterate this process, we obtain sequences (a_n) and (b_n) defined by

$$a_0 = a \qquad\qquad b_0 = b$$
(8)
$$a_{n+1} = (a_n+b_n)/2 \qquad\qquad b_{n+1} = (a_n b_n)^{1/2}.$$

The   above inequality enables one to show easily that these sequences converge
to   a   common   limit   M(a,b)   which we define to be the   agM   of   a   and   b
([3],[4] and [5]).

    The   usefulness   of the formula (7) for the calculation of   K(k)   appears
evident on the computer, the algorithm that defines M(a,b) being rapidly
convergent.

    In the previous exercise the argument is particularly suited to numerical
elaboration and computer application. It is obvious that for the treatment of
other   arguments   traditional   theoretical   lessons   satisfy completely   the
didactical aims, while for other arguments the use of ready-made software can
still be useful.

    Our didactical research project, in conclusion, can be identified with a
selection of experiments, in wich the actors are the instructors and the
students while the scenery is both the traditional background of textbooks,
chalk and blackboard lessons and the innovative one of instruments and
products of computer science.

REFERENCES

[1] Boieri, P. and oth., Personal computers in teaching basic mathematical courses, SEFI Erlangen (1984).
[2] Buzano, P., Lezioni di Matematica per allievi ingegneri, n. 3 (Levrotto & Bella, Torino, 1976).
[3] Cox, D., The arithmetic-geometric mean of Gauss, L'Enseignement Math. 30 (1984) 275.
[4] -------, Gauss and the Arithmetic-Geometric Mean, Notices AMS 240 (1985) 147.
[5] Gauss, C.F., Werke (Göttingen-Leipzig, 1868-1927).
[6] Mascarello, M. and Scarafiotti, A.R., Computer experiments on Mathematical Analysis teaching at the Politecnico of Torino, in Supporting papers (ICMI Strasbourg 1985) 265.
[7] Mascarello,M. and Winkelmann, B., Calculus and the computer. The interplay of discrete numerical methods and calculus in the education of users of Mathematics: considerations and experiences, in: Kahane, J.P. and Howson, A.G., (eds) The influence of computers and informatics on mathematics and its teaching, (Cambridge University Press, Cambridge, 1986) 120.
[8] Smirnov, V.I., Corso di matematica superiore, n. 2 (MIR, Mosca, 1976).
[9] Tricomi, F.G., Funzioni ellittiche (Zanichelli, Bologna, 1951).

ECM/87 - Educational Computing in Mathematics
T.F. Banchoff et al (editors)
© Elsevier Science Publishers B.V. (North-Holland), 1988

The Mandelbrot Set:
A Paradigm for Experimental Mathematics

Heinz-Otto Peitgen*, Hartmut Jürgens and Dietmar Saupe

Institut für Dynamische Systeme, Universität Bremen

2800 Bremen 33, FR Germany

Abstract. Since the discovery of the **Mandelbrot set** in 1980, the scienes have been enriched by one of the most complicated and beautiful objects which they have ever seen. Its complexity and richness of structures and forms and the depth of mathematical problems connected with these is in amazing contrast to the simplicity of its generating code and needed the power of modern computer graphics to be revealed.

A typical code for the generation of the Mandelbrot set needs only a few lines based on the simple dynamical system

$$x_{n+1} = (x_n)^2 + c, \quad n=0,1,2,\ldots$$

(1)

$$x_n \in \mathbf{C}$$

in the complex plane $\mathbf{C}$. It exemplifies in a most beautiful way the power of experimental and computational methods in higher education. The Mandelbrot set, figure 1, embodies several important paradigmatic characterizations for the complex dynamical behavior of (1). For example, it is the **bifurcation set** of periodic orbits of (1) and describes various routes from **order** into **chaos**. Our interest is in its interpretations for a **morphology of fractal basin boundaries**, known as **Julia sets**. For example, values of c on the boundary of the Mandelbrot set locate a crisis in the associated Julia set of (1). The crisis is that for c-values from the Mandelbrot the associated Julia sets are connected, while for c-values outside the Mandelbrot set they are totally disconnected (a cloud of infinitely many points).

---

* Department of Mathematics, University of California, Santa Cruz, CA 95064, USA

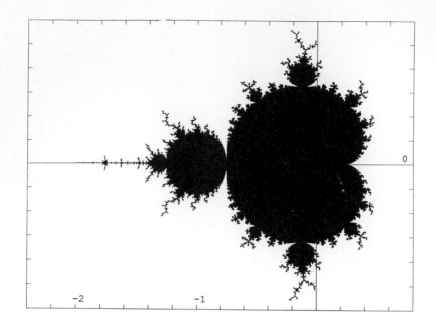

Figure 1. The Mandelbrot Set

Formally one defines the Julia set for $f_c(x) = x^2 + c$ as follows: Fix $c \in \mathbf{C}$. Then let K_c be the set $\{x_0 \in \mathbf{C} : x_n$ remains bounded for all $n\}$, where x_n is obtained by (1) . Since all finite periodic points are in K_c one has that $K_c \neq \emptyset$. The complement of K_c in $\Sigma = \mathbf{C} \cup \{\infty\}$ is denoted by $A_c(\infty)$ and can be interpreted as the basin of attraction of the attractive fixed point ∞ of f_c, i.e. $A_c(\infty) = \{x_0 \in \mathbf{C} : x_n \to \infty$ with $n \to \infty \}$. Now the Julia set of f_c is defined as $J_c = \partial K_c = \partial A_c(\infty)$.

It is known from the classical work of G.Julia and P.Fatou that the J_c's obey a simple dichotomy: J_c is

- either connected,
- or totally disconnected.

Figure 2 shows a collection of a few Julia sets of both types. For example, figure 2e is obtained for c = i; figures 2d and 2f are disconnected and all other Julia sets in figure 2 are connected. This is , however, a rather incomplete list. In fact, there are infinitely many significantly different Julia sets, namely one for each choice of c in (1). The Mandelbrot set

(2) $M = \{ c \in \mathbf{C} : J_c$ is connected$\}$

can be seen as an order-principle in that infinite variety of J_c's. It is not hard to show that the dynamical behavior of the critical point c determines M numerically

(3) $M = \{ c \in \mathbf{C} : f_c^k(0)$ remains bounded for all k$\}$.

Here f_c^k denotes the k-th iterate of f_c, i.e. $f_c^k = f_c (f_c^{k-1})$.

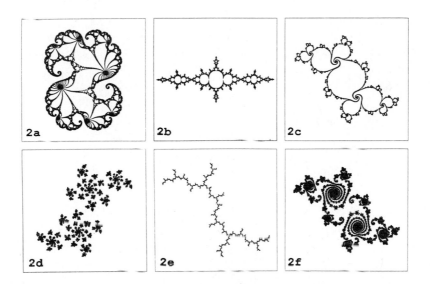

Figure 2. Collection of Julia Sets

Figure 2.(continued)   **2g**

On of the most striking interpretations of the Mandelbrot set is that it can be viewed as a **"one picture dictionary"** of these infinitely many shapes of Julia sets. This gives some flavor of its almost unimaginable complexity which still would be in the dark without the advent of modern computer graphics. This note is devoted to an intuitive discussion of this amazing **"image compression"**.

Our first experiment in figure 3 illustrates a beautiful result of Tan-Lei [4]. A region along the cardioid of M (see (a) in figure 3) is continuously blown up and stretched out, so that the respective segment of the cardioid becomes a straight line segment. The result is shown in (b). The effect of this particular blow up is that now each member of a sequence of certain satellites (colored solid black) has the same size.[1] Attached to the satellites one recognizes **dendrite structures** with $3,4,\ldots,10$ major branches (from top to bottom). Choosing as c-values the **major branch points** of these structures, which are examples of Misiurewicz points, (see for example arrow in the top dendrite) one obtains 8 Julia sets which are shown on the right of (b). Remarkably, our blow up of the Mandelbrot set reflects certain structural and combinatorial aspects of these Julia sets in an amazing correspondence, which is not accidental. In fact this phantastic property of the Mandelbrot set is true everywhere along its boundary . More precisely, it is true for a dense subset of the boundary of M - the set of Misiurewicz points.

A point c is called a Misiurewicz point provided it is preperiodic. I.e. $f_c^k(c) = c^*$, $k > 0$, and $f_c^m(c^*) = c^*$ for some $m > 0$. In other words, c is a Misiurewicz point provided the orbit of 0 under f_c is eventually periodic but not periodic. If m is the minimal period of c^* the number $\rho = (f_c^m)'(c^*)$ is called the eigenvalue of the m-cycle c^*. It is known that $|\rho| > 1$. Now Tan-Lei's result is:

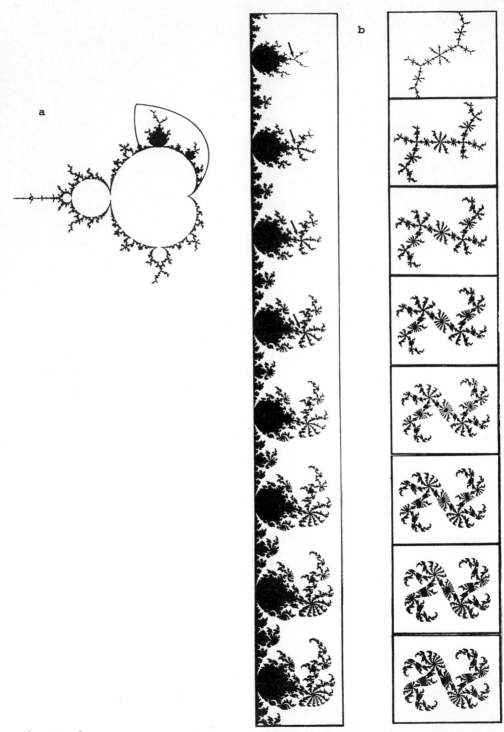

Figure 3.

THEOREM ([4])

Let c be a Misiurewicz point and ρ its eigenvalue. Then there is $\lambda \in \mathbf{C}-\{0\}$ and a closed subset Z in $\mathbf{C}$ such that

- $\rho Z = Z$

- $\rho^n(\tau_{-c}(K_c)) \rightarrow Z$

- $\rho^n(\tau_{-c}(M)) \rightarrow \lambda Z$.

(τ_{-c} denotes the translation by -c and "$\rightarrow$" is convergence in a suitably modified Hausdorff metric.)

The theorem can be interpreted in the following way: Suppose one magnifies the Mandelbrot set around a Misiurewicz point c and compares with a magnification around the point c in K_c by the same blow up factor. Then in a small window the magnifications will be essentially indistinguishable, provided the blow up factor is large. But there is more and this will be discussed in our next experiment. For that let c = i. Then $f_i(i) = -1+i$, $f_i(-1+i) = -i$ and $f_i(-i) = -1+i$. i.e. c = i is a Misiurewicz point. Moreover $\rho = (f_i^2)'(-i) = \sqrt{32}\ \exp(2\pi i/8)$. Figure 4 shows 12 blow ups of M around c = i. The blow up factor from one to the next is $\sqrt{32}$. We observe an almost perfect selfsimilarity as we go from a to l (compare figure 4d with 4l). Moreover the tip of the dendrite at c = i just appears to be rotated by 45^0 in the subsequent blow ups from d to l, so that after a blow by 32^4 the tip of the dendrite just appears as if it had been rotated by 360^0. In other words, the tip of the dendrite is actually a spiral, as it should be according to $\rho Z = Z$ in Tan-Lei's result. This fact comes as a surprise for those who believed to be somewhat acquainted with M based on computergraphical studies.

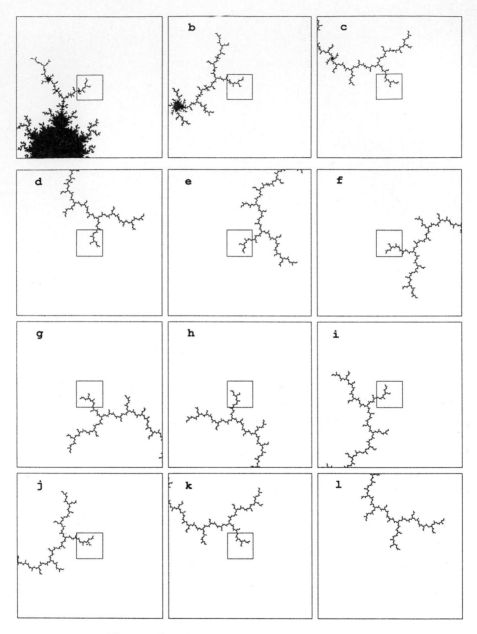

Figure 4. Blow ups of M around c = i.

The selfsimilarity seen at Misiurewicz points is somehow misleading in regard to the complexity which one finds if one zooms in onto the Mandelbrot set. Figure 5 shows a sequence of 7 successive magnifications at the boundary of M. The window for each blow up is indicated by a frame. Though images of M and its blow ups could be obtained from the numerical characterization (3) in principle, this method is usually not a suitable graphical method (see [2,3] for a detailed discussion of graphical methods for complex dynamical systems). Our experiments in figure 5 use a method which estimates the euclidean distance d(c,M) of a point c outside of M to the boundary of M. This method is based on an estimate by J.Milnor which claims that

(4) $d(c,M) \leq 2 \sinh(G(c))/|G'(c)|,$

where $G:\mathbf{C}\text{-}M \rightarrow \mathbf{R_+}$ is the potential of M. Using this and the representation

(5) $G(c) = \lim_{k\to\infty} 2^{-k} \log(|f_c{}^k(0)|)$

one derives an efficient code for the graphical representation of the Mandelbrot set (see[3])[2].

Our final experiment (in color) discusses the similarity between the Mandelbrot set and Julia sets for points in the interior of M. Figure 5f shows a detail of M, the center of which reveals a miniature copy of a Mandelbrot set. Choosing c from the center of this miniature copy the dynamical system (1) has an attractive periodic cycle of order 45. Figure 6(color)[*] shows a global view of the Mandelbrot set and color figure 7(color)[*] shows the blow up, which is identified by the yellow dot in color plate figure 6. The magnification here is identical with figure 5f though the method of its generation is quite different[3].

---

* See page 284 for Figures 6 and 7.

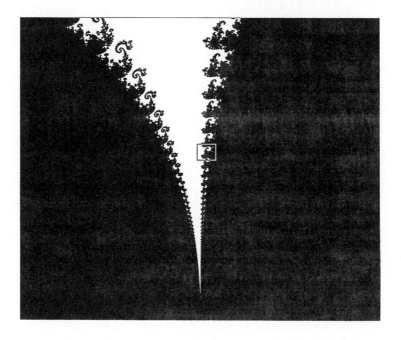

Figure 5.Successive Blow Ups at the Boundary of Mandelbrot Set

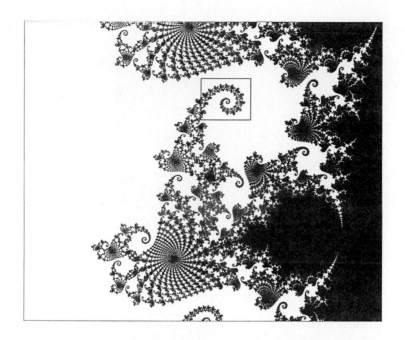

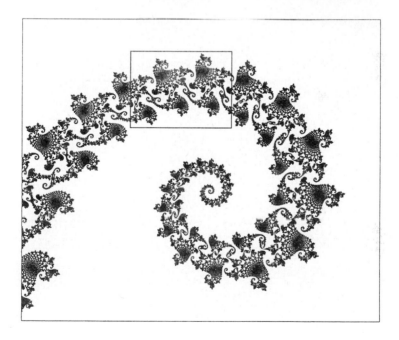

Figure 5. (continued)

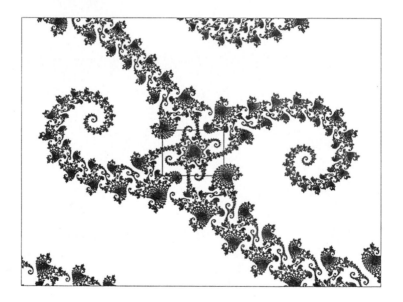

Figure 5. (continued)

Figure 5. (continued)

Figures 8[*] and 9[*] (color) show the Julia set and a blow up corresponding to the c-value from the center of figure 7[*] (color). The yellow dot in figure 8[*] (color) identifies the magnification shown in figure 9[*] (color). The dot marks a window centered at c. The magnification factor for 4 is the same as for figure 7[*] (color). However, the structure in figure 9[*] (color) is shown after a rotation of approximately 55^0. The similarity of figures 7[*] and 9[*] (color) is striking and is subject of current research.

The true complexity of these experiments can only be revealed using color graphics, as for example in [2].

[1]) The satellites correspond to stability regions of certain periodic cycles of (1), i.e. for c values from a particular satellite the dynamical system (1) admits an attractive cycle of some fixed order; from top to bottom of figure 3 (II) these are stable 3-, 4-, ...,10-cycles. The magnification factor is based on a conjecture, which says that the areas of the satellites along the cardioid decay like $1/p^2$ where p is the period identified by a satellite.

[2]) Distance Estimator Method for M: Choose N_{max} (maximal number of iterations) and R large, e.g. R = 100.

For each   c   one determines a label l(c) from {0, 1, 2}   (0 for c ∈ M , 1 for c close to M, 2 for c not close to M) :
Compute

$$x_{k+1} = (x_k)^2 + c , \quad x_0 = 0 , \quad k = 0, 1, 2, \ldots$$
$$\text{until either} \quad |x_{k+1}| \geq R \quad (\textbf{then} \quad \textbf{l(c)} = \textbf{2})$$
$$\text{or} \quad k = N_{max} \quad (\textbf{then} \quad \textbf{l(c)} = \textbf{0})$$

If   **l(c)** = **2** , then   c   is still a candidate for a point close to M. Thus, we try to estimate its distance, having saved the orbit $\{x_0, x_1, \ldots, x_n\}$, where   n   is the first index such that $|x_n| \geq R$:

$$x_{k+1}' = 2x_k x_k' + 1 , \quad x_0' = 0 , \quad k = 0,1, \ldots , n$$
(6)
$$\text{if} \quad |x_{k+1}'| = \text{OVERFLOW} , \quad \textbf{then l(c)} = \textbf{1} .$$

If in the course of the iteration (6) we get an overflow, then   c   should be **very** close to M , thus we label   c   by 1.
Finally, estimate   the distance of   c   from M :

$$\text{if } (|x_n|/|x_n'|) \log (|x_n|) < \text{DELTA} , \quad \textbf{then} \quad \textbf{l(c)} = \textbf{1}$$

$$\text{otherwise} \quad \textbf{l(c)} = \textbf{2}$$

It turns out that the images depend very sensitively on the various choices (R, N_{max}, OVERFLOW, DELTA and blow-up factor when magnifying M). Therefore, it is appropriate to extend the labeling function   by estimating points which have a distance given by a fraction of DELTA (rDELTA/s , r = 1, ... , s).

---

[*] See page 284 for Figures 6, 7, 8 and 9.

Using a proper color saturation for each of these extended labels usually gives very satisfactory results.

3) <u>Level Set Method for M</u>: Fix a square lattice of pixels in the c-plane, choose a large integer N_{max} (iteration resolution) and an arbitrary set T (target set) containing ∞. For example, $T = \{c : |c| \geq 1/\varepsilon \}$, – ε small – is a disc around ∞ . Now we assign for each pixel p from the lattice an integer label $L(p;T)$ in the following way: (p identifies a number c, e.g. the center of p)

$$L(p;T) = \begin{cases} k, \text{ provided } \quad f_c^{\,k}(0) \in T \quad \text{and } 1 \leq k \leq N_{max} \text{ and } f_c^{\,i}(0) \notin T \\ \qquad\qquad\qquad\qquad\qquad\qquad\qquad\qquad \text{for } 0 \leq i < k \,. \\ 0, \text{ else.} \end{cases}$$

The interpretation of $L(p;T) \geq 1$ is obvious: p (or rather the value c = $f_c(0)$ which p represents) escapes to ∞ under the iteration of f_c and $L(p;T)$ is the "escape time" - measured in the number of iterations - needed to hit the target set around ∞ . The collection of points of a fixed label, say k , constitutes a level set, the boundary of which is the union of two circle-like curves, provided T is the complement of a large disc.

References

[1] Blanchard, P.: Complex analytic dynamics on the Riemann sphere, Bull.Amer.Math.Soc.11(1984),85-141.

[2] Peitgen, H.O., Richter, P.H.: The Beauty of Fractals, (Springer-Verlag, Heidelberg, 1986).

[3] Peitgen, H.O. (ed.): The Science of Fractals, (Springer-Verlag, New York, 1988).

[4] Tan-Lei:Ressemblance entre l'ensemble de Mandelbrot et l'ensemble de Julia au voisinage d'un point de Misiurewicz,preprint.

ECM/87 - Educational Computing in Mathematics
T.F. Banchoff et al. (editors)
© Elsevier Science Publishers B.V. (North-Holland), 1988

CALCULUS WITH A PC

Fred Simons

Department of Mathematics and Computing Science,
Eindhoven University of Technology, P.O. Box 513,
5600 MB Eindhoven, The Netherlands.

The programs on the diskette accompany the book "Calculus with a PC" [1], but can be used independently. They are written in GWBASIC for IBM-compatible personal computers. The diskette also contains the ASCII-files of the programs. Therefore, the programs can easily be adapted to the computer of the user by using a word processor.

The programs are designed both for obtaining numerical or graphical results and for illustrating some of the concepts in a first year Calculus course.

Usually, the numerical output of a computer is the result of approximations and an approximation is of no use if no information is available on the accuracy of the approximation. Some numerical procedures, for example the bisection method, actually produce an interval containing the wanted number, and then the question about the accuracy is easily answered. However, for other procedures, such as Simpson's rule for approximating a definite integral, it is much more difficult to get an idea about the accuracy. Often the accuracy can be estimated by means of analytical techniques, but for the students this usually means a lot of tricky and technical computations.

Therefore, in the programs information about the accuracy is obtained in a numerical way. Suppose that we want to approximate the value of a quantity α, for example a definite integral. We then construct a sequence (a_n) that converges at least linearly to α and we estimate the value of α by the ultimate constant decimal digits in the sequence. To be sure that the sequence indeed converges at least linearly, the sequence of successive differences and the sequence of quotients of successive differences is also computed. The latter sequence should converge to the convergence factor and therefore should have a limit between -1 and 1. For details the reader is referred to [1], section 6.1.

As an example we show how we compute the value of π by means of numerical integration.

$$\pi = \int_0^1 \frac{4}{1 + x^2} \, dt$$

The following table shows the sequence of approximations with the midpoint rule from the program NUMINT.

n	value	difference	quotient
1	3.200000000		
2	3.162352941	-0.037647059	
4	3.146800518	-0.015552423	0.4131
8	3.142894730	-0.003905789	0.2511
16	3.141918174	-0.000976555	0.2500
32	3.141674034	-0.000244141	0.2500
64	3.141612999	-0.000061035	0.2500
128	3.141597740	-0.000015259	0.2500
256	3.141593925	-0.000003815	0.2500
512	3.141592971	-0.000000954	0.2500
1024	3.141592733	-0.000000238	0.2500

Indeed, the convergence factor is 0.2500, which is the theoretical value as well, and therefore we conclude that the sequence convergences linearly and

$$\pi = 3.141\ 59 \pm 0.000\ 01\ .$$

Here is another example. We consider the sequence $(1 + \frac{1}{n})^n$. Its limit is the number e. With the program sequence we obtain the following output:

n	term	difference	quotient
131	2.7079787813	0.0000786996	0.9848
132	2.7080562967	0.0000775154	0.9850
133	2.7081326544	0.0000763577	0.9851
134	2.7082078802	0.0000752258	0.9852
135	2.7082819991	0.0000741188	0.9853
136	2.7083550352	0.0000730362	0.9854
137	2.7084270122	0.0000719770	0.9855
138	2.7084979530	0.0000709407	0.9856
139	2.7085678797	0.0000699267	0.9857
140	2.7086368139	0.0000689342	0.9858
141	2.7087047767	0.0000679628	0.9859

It appears that the sequence does not converge linearly and therefore we cannot draw a conclusion on the value of e from the table.

Now let us discuss the programs on the diskette.

ONEVAR is a program for analyzing a function of one variable. The options are to evaluate the function at a given point, to make a table of values, to approximate the zeros and the extreme values and to plot the graph on the screen. The approximation of a zero uses the bisection method and the approximation of an extreme value uses the "golden section search". Therefore the derivative of the function is not needed for finding the extreme values.

CURVE is a similar program for curves in the plane. De curve is given by its two coordinate functions. The options are to make a table, to find the zeros and the extreme values of each of the coordinates, to plot in one figure each of the two coordinate functions and to plot the curve itself.

SEQUENCE is a program for analyzing the convergence of a sequence. It produces the elements of the sequence, the sequence of successive differences and the sequence of quotients of successive differences, the latter providing us the convergence factor of the sequence.

SERIES is a similar program for series. The sequence of partial sums, the sequence of the terms and the sequence of the quotients of successive terms are produced. The latter sequence has the rate of convergence of the sequence of the partial sums as its limit.

SUCSUB deals with sequences given by the relation $a_{n+1} = f(a_n)$ for a given initial value a_1. It is similar to the program SEQUENCE, but also shows the geometrical construction of the sequence. Figure 1 shows a screendump.

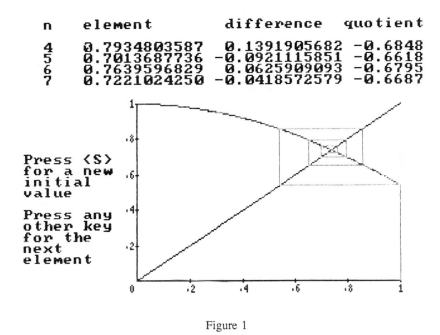

Figure 1

ZERO contains some methods other than the bisection method used in ONEVAR for approximating a zero of a function: the regula falsi, the secant method, Newton's method and the method of fixed slope.

NUMINT is a program for numerical integration. The midpoint rule, the trapezium rule and Simpson's rule can be used and compared.

DIFFEQ can be used both for numerical and for illustrative purposes. It can plot the direction field and a solution of a differential equation, and it can approximate the solution of an initial value problem numerically with either of the methods of Euler, Euler-Heun, Euler-Cauchy and the fourth order Runge-Kutta method. The following figure shows an example.

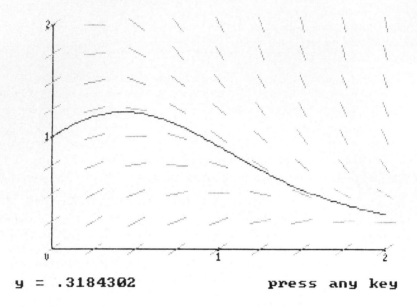

y = .3184302   press any key

Figure 2

TAYLOR is a program only for illustrative purposes. With this program, the graph of a function together with some of its Taylor polynomials can be plotted.

Finally, with the program FOURIER we can plot the graph of a periodic function together with some of the partial sums of the Fourier series. In this program, the Fourier coefficients can either be entered manually, or be approximated by means of numerical integration.

References
[1] Simons, Fred, Calculus with a PC (McGraw-Hill, Hamburg, 1987).

ECM/87 - Educational Computing in Mathematics
T.F. Banchoff et al. (editors)
© Elsevier Science Publishers B.V. (North-Holland), 1988

AN ALGORITHM FOR RECURSIVE GENERATION OF A CLASS
OF CONVEX UNIFORM POLYHEDRA

A. Spizzichino, TESRE/CNR, via de' Castagnoli, 1, Bologna, Italy

In order to generate a data base as wide as possible concerning regular or semiregular polyhedra, starting from a minimal set of data, an algorithm for constructing convex uniform polyhedra of the topologically simplest type (connection order = 3) has been developed. The class includes 3 Platonic solids, 7 Archimedean solids (all the 5 truncated and the 2 rumbitruncated) and, of course, the infinite set of the Archimedean prisms. The only input required is the number of sides of the three faces meeting at one vertex (the Cundy-Rollett symbol) from which the triplet of concurrent faces is generated and its boundary extracted (seed boundary). The solution is achieved by regarding the polyhedron as a growing structure, in which the accretion process is strictly related to the seed and terminates when the boundary degenerates to a point. A full description of the construction method is given, as well as of the manner in which each subsequent accretion boundary is derived. The way of approaching the problem may be of educational interest and lead to applications in related areas of computational geometry.

1. INTRODUCTION

Convex uniform polyhedra (CUP) in three dimensional space may be regarded as the solution of the classical problem of wrapping a sphere with an equilateral flexible network. It is well known that the problem has 18 different solutions (not including the infinite set of prisms and anti-prisms), 5 of which, characterized by having identical meshes, correspond to the Platonic solids, while the remaining 13, having meshes of different types, correspond to the Archimedean ones.

Several representations (or symbols) have been proposed for CUP's description and identification [1-4]. Both from a practical and an educational point of view it would be useful to set up an algorithm capable of deriving, for each CUP, the coordinates of the vertices and the topology of their connections, starting from their bare symbolic representation.

Indeed, the non-specialist who wants to produce an image of a particular regular or semi-regular polyhedron making use for instance of the tools of computer graphics, is heavily hampered by the lack of a data base with this information, in spite of the circumstance that the symbol itself contains it in an implicit way.

The problem stated in this paper can then be summarised as follows: how to derive the data base concerning a given CUP from the descriptive properties contained in its symbol, or, in other words, from the bare

A. Spizzichino

inspection around one vertex? A general algorithmic approach to this question has not been achieved, but, as a partial solution, a simple computational procedure, valid for Archimedean prisms and for 10 out of the 18 CUP is presented in the following.

2. THE ALGORITHM

The subset of CUP considered here is that characterized by a 3-connected network. We shall refer to it as CUP_3 (Table 1).

Table 1

3-CONNECTED CONVEX UNIFORM POLYHEDRA

Platonic solids	Symbol
1 - tetrahedron	(3.3.3)
2 - hexahedron	(4.4.4)
3 - dodecahedron	(5.5.5)
Archimedean solids	
4 - truncated tetrahedron	(3.6.6)
5 - truncated octahedron	(4.6.6)
6 - truncated hexahedron	(3.8.8)
7 - truncated icosahedron	(5.6.6)
8 - truncated dodecahedron	(3.10.10)
9 - rumbitruncated cuboctahedron	(4.6.8)
10 - rumbitruncated icosidodecahedron	(4.6.10)

Let us consider a connected portion of a CUP_3 surface made up of an integer number (>2) of faces, along with the skew equilateral polygon representing its boundary (Figure 1). Exploring the boundary by pairs of contiguous edges, we observe that only if the couple is heterogeneous, that is only if it belongs to two different faces, does it generate a new face of the polyhedral surface. Let P_2 be the common point of the two edges of an arbitrary heterogeneous couple P_1 P_2, P_2 P_3. We generate the new face as the regular polygon having vertices at P_1, P_2, P_3, whereupon we redefine the new boundary by elimination of the common edges. This procedure will be iterated until the boundary degenerates into a point, thereby indicating the completion of the construction.

The generation of the initial minimum (or "seed") boundary can be described as follows. Referring to Figure 2, let us consider two contiguous faces f_1 and f_2 assuming that their common edge lies on the x axis. Let U, O and V be the first three vertices of f_1, and W, O and U the first three

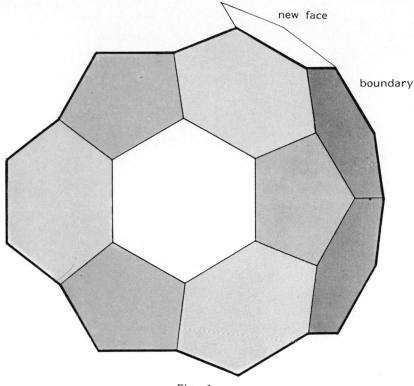

Fig. 1

vertices of f_2. Assuming edges of unit length, in vector notation we can write

$$\bar{u} = (1,0,0)$$

$$\bar{v} = (\cos \theta_1, -\sin \theta_1, 0)$$

Then $\bar{w}$ is obtained from the vector equations

$$\bar{u} \cdot \bar{w} = \cos \theta_2$$

$$\bar{v} \cdot \bar{w} = \cos \theta_3$$

$$\bar{w} \cdot \bar{w} = 1$$

which, when solved, give the components

$$w_1 = \cos \theta_2$$

$$w_2 = (\cos \theta_3 - v_1 w_1)/v_2$$

$$w_3 = (1 - w_1^2 - w_2^2)^{1/2}$$

After the triplet $\bar{u}$, $\bar{v}$ and $\bar{w}$ has been defined, the seed boundary is easily obtained eliminating the edge OU from the union of the boundaries of f_1 and f_2.

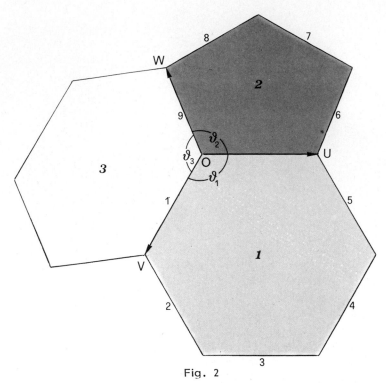

Fig. 2

As we require the polyhedron to be centred on the origin, the centre C of the circumscribed sphere must be translated to O, and the same translation vector $(-\bar{c})$ must be applied to all the vertices.

The components of $\bar{c}$ are derived requiring that the sphere

$$x^2 + y^2 + z^2 + \alpha x + \beta y + \gamma z + \delta = 0$$

passes through O, U, V and W, and recalling that its centre C is given by

$$C = (-\alpha/2, -\beta/2, -\gamma/2)$$

After a little algebra, in our case we obtain

$$\alpha = -1$$

$$\beta = (v_1 - 1)/v_2$$

$$\gamma = (w_1 - \beta w_2 - 1)/w_3$$

It is clear from the above broad line description that the only input needed to produce the seed boundary and the entire CUP_3 data structure is the sequence of the number of sides of the polygons meeting at one vertex, that is to say the symbol used by Cundy and Rollett, a minimal data set from which the vertex angles, the edge triplet $\bar{u}$, $\bar{v}$, $\bar{w}$ and all the needed related quantities can be derived.

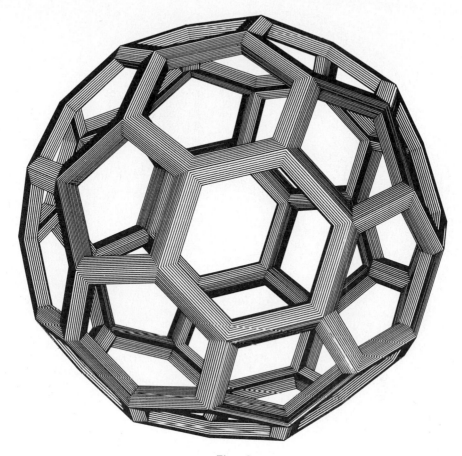

Fig. 3

A program to produce CUP_3 data sets using this algorithm was written in Basic and implemented on an Apple II personal computer. From these we have been able to reproduce a collection of polyhedron stick models in the style of the Renaissance artists, two examples of which are shown in Figures 3 and 4. Although the latter program is necessarily quite elaborate, it can still be run on an Apple II.

3. CONCLUSIONS

The modest computational requirements of the algorithm presented here mean that, even on a small machine, after careful implementation the time taken to produce a CUP_3 dat set is practically reduced to the output stage alone.

This procedure, avoiding the entering of cumbersome amounts of data, is therefore ideally suited for use in fields such as architectural design, college

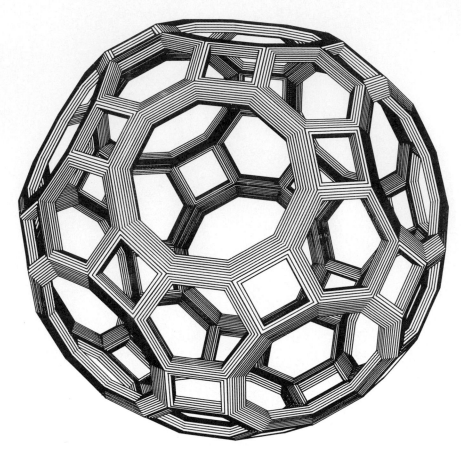

Fig. 4

maths, crystallography, etc. A particular feature of its approach is that instead of relying on symmetry, truncation or duality, it employes uniformity and connectivity concepts, as suggested by many natural growth structure.

    If it is true in general that computers can make geometry more attractive for students and teachers, it is also true that a lot of work has to be done to realise this potential. Hopefully, this paper is a contribution towards this goal.

REFERENCES

[1] Cundy, H.M., and Rollett, A.P., Mathematical Models (Oxford University Press, London, 1961).
[2] Coxeter, H.S.M., Regular Polytopes (Dover, New York, 1963).
[3] Wenninger, M.J., Polyhedron Models (Cambridge University Press, Cambridge, 1971).
[4] Williams, R., The Geometrical Foundation of Natural Structure (Dover, New York, 1972).

2. TEACHING EXPERIMENTS

ECM/87 - Educational Computing in Mathematics
T.F. Banchoff et al. (editors)
Elsevier Science Publishers B.V. (North-Holland), 1988

GALOIS: AN INSTRUCTIONAL SYSTEM IN ALGEBRA
FOR COMPUTER SCIENCE STUDENTS

M.A.Alberti, A.Bertoni, G.Mauri, N.Sabadini, M.Torelli

Dipartimento di Scienze dell'Informazione
Università degli Studi di Milano
Via Moretto da Brescia 9
I 20133 Milano, Italia *

An interactive educational system on algebra for computer science students
is described. The system generates problems at two different level of
difficulty. An introductory module provides the basic definitions and
concepts. At each exercise explanations can be visualized. The ideas
underlying the system and the educational background are described.

1. INTRODUCTION

GALOIS is an interactive system designed for students in Computer Science, who have to
master a first course in algebra. The system has been developed in connection with one of the
algebra courses offered at the University of Milan. The system is intended to be a learning tool,
which the students can benefit from, besides class lectures and exercises hours. The students are
not encouraged to work on the system without having read throughout the course book [2] or
another text book, for example [7].

The users may select topics of their choice and solve problems, generated by the system, on
that topic. The problems to be solved require the students to learn to use basic manipulation
techniques and algorithms to derive numerical or yes/no answers. The system does not ask to
prove properties or theorems since it has no automatic theorem prover to check the correctness of
the answers.

Testing is an important learning mode, that can eventually encourage practice and
understanding, but we also think that repetition of exercises rarely leads to learning of the
underlying concepts and principles. After some understanding of the concepts and principles
involved in the problem has occurred, then testing and practising give deeper insight and more
confidence in them.

The users have control over the subject and the pacing of the system; they can use it to
check the progress of their learning or to practice before taking the examination. No records are
kept of the student-system interaction, because the system is not meant as an evaluation tool for
the teachers.

The system generates problems at two different levels of difficulty, according to the user's
choice; explanations can be visualized for each problem. On the other hand, some topics are
treated at an introductory level; in this case the user can read definitions and simple concepts and

* Work financially supported by Ministero della Pubblica Istruzione (funds MPI 60%)

can ask for problems if he/she feels the need of practising. That is, in the presentation of introductory notions the stress is more on tutoring than on problem solving. Whereas in the other cases the interaction is taking place exactly the other way round: users solve problems and they can ask for explanations. The reason for this is that the introductory module is mainly intended to make the use of terminology uniform and recall simple definitions.

2. EDUCATIONAL FRAMEWORK

Algebra is an important subject for its many and wide applications in different fields; therefore there is no need to justify its introduction in a university curriculum. Here we want to emphasize that some of the ideas of modern algebra are applicable to computing and that notions as those of algebraic structures and classes of isomorphisms are central for a computer science curriculum.

Let us think for instance of the role played in programming languages by the notion of abstract data type [5] [8]. In the algebraic approach to abstract data types, a data type is defined as a set of operation symbols and constant symbols and a set of equations among terms (with variables), specifying the operations properties. Hence, a *canonical*, unique (up to isomorphisms) model is constructed. Defining a set of data objects and operations on them according to the given properties, one obtains another model, related to the canonical one by a morphism, which is an implementation of the abstract data type.

Moreover, for computer science students it is important to get acquainted with topics in logic and discrete and combinatorial mathematics, which are also in the course syllabus. Further readings on these topics are found in [1] [3] [4] [11].

But perhaps the most important educational aspect of an algebra course is to allow the students to become familiar with concepts at different levels of abstraction in a smooth progression: from the *concrete* level of mathematical objects as integers, polynomials, matrices and their operations to the more abstract level of the algebraic structures defined by postulates as rings, fields and so on, up to the more general concepts of universal algebra. Moreover there is a very nice correspondence between the escalation from the intuitive concepts of set and subset to that of function and equivalence relation, and, on the side of algebraic systems from the concepts of algebra and subalgebra to that one of morphism and congruence.

Therefore the goals of the GALOIS system, which follows the general lines of the parallel course, are in general to increase students' abstraction ability while providing them basic mathematical knowledge and could be summarized as follows:

- ◻ to develop the students' ability to separate the abstract notion of structure from its possible representations;
- ◻ to develop the students' problem solving ability, in particular by asking them to solve problems on basic algebraic structures;
- ◻ to help the students   become familiar with different algebraic and combinatorial structures; for instance the system includes some simple counting problems, which are rather unusual in traditional algebra exercises.

At the first level of difficulty basic operations on algebraic structures are introduced. This requires learning basic algorithms and manipulative techniques. At the second level of difficulty the students are asked to find the correct sequence of basic operations and/or use some general principles to solve the more complex and *structured* problems. Students have first to acquire a concrete manipulative experience on algebraic structures to be able to operate at a higher level of abstraction.

Let us make this point clear with an example. At the first level, among the problems on graphs we find the following exercise: "given an oriented graph, determine the number of paths of length n between two given vertices". This problem requires, besides basic knowledge on graphs, the ability to use different representations of the notion of relation (graph and adjacency matrix) and elementary operations of matrix algebra. At the second level the corresponding exercise requires to determine the number of paths which satisfy certain constraints (e.g. paths passing through a given vertex and not passing through another given one). This exercise can be solved using the inclusion-exclusion principle.

Another idea underlying the system is that algorithm schemes allow to solve problems on different algebraic structures in a rather uniform way. Recent symbolic manipulation systems are based on this idea too [6].

As an example: Euclid's algorithm to compute the gcd of integers can be extended to general Euclidean domains, thereafter allowing to solve, for instance, Diophantine equations on polynomials.

3. CONTENTS OF THE SYSTEM

The topics covered by the system, which are not exhaustive of the whole syllabus, follow together with a brief explanation.

Preliminary notions in mathematical logic

The languages of propositions and predicates. Interpretation and truth value of a formula.

Sets, relations and functions

Sets and operations on them. Different types of relations and functions and their enumeration (a topic not usually covered in algebra textbooks). Use of the adjacency matrix to decide properties of relations. Construction of relations with given properties.

Graphs and paths

Graphs and their adjacency matrices. Enumeration of paths with prescribed constraints.

Modular arithmetic, continued fraction decomposition,

Diophantine equations and Chinese Remainder Theorem

Solution of linear Diophantine equations and of systems of modular equations with coprime moduli.

Algebraic structures, isomorphisms

Mainly cyclic groups and finite fields; substructures and congruences; isomorphisms.

Polynomial operations and irreducibility

Sum, product, division and gcd of polynomials with coefficients in a given field. Continued fraction decomposition of polynomials. Solution of linear equations. Conditions for the irreducibility of a polynomial in a given field.

4. ARCHITECTURE AND IMPLEMENTATION

The system includes different modules taking care of user interface, problem generation, solution computation, explanations displaying and answers checking.

As for the interaction with the user, the selection of topics is carried out through menus at different levels of detail. The displaying of successive menus provides a list of contents of the system. The first choice the user is asked to make is among introductory, elementary and more complex exercises. An elementary problem is thought of as one which can be solved by knowing just that topic, besides of course standard high school algebra. Difficult problems are those which conceptually imply one or more elementary exercises.

At each problem the user has three chances. He/she can give an answer to the exercise, or escape back to the previous menu, or ask for explanations. When the user requires explanation of a difficult exercise, the system provides easier exercises to practise. Figure 1 on next page describes the details of the relation between difficult and easy exercises by means of Petri nets [9], which are a good tool for modeling the behaviour of discrete dynamic systems. A description of the net can be found in the next section.

4.1 Problem generation

The different problems are generated at random within a certain module at the moment the student calls for them. Each problem is associated with a fixed scheme and some parameters, and with an algorithm that computes the solution (or the solutions if more than one).

For some problems the parameters are just numbers: randomly generated within a certain range (e.g. coefficients and degrees of polynomials or matrix elements), or generated so that they satisfy given constraints (e.g. coefficients that must be coprime in some Diophantine equations). For other problems parameters are the arcs connecting vertices in a graph or even a particular sentence chosen among a few that characterize the problem (e.g. "determine the number of paths of length x connecting vertex a to vertex b and *sentence*" . Here x, a, b are integer parameters and *sentence* is generated among some 15 different ones such as: "passing through c and not through d" or "if not passing through c then passing through d").

In this way we can guarantee some variety to the problems while preventing awkward cases and ensuring the existence of the solution. Sometimes the solution to a problem is randomly generated and afterwards the problem admitting such a solution is computed (e.g. in the problem: "given a graph find its adjacency matrix" at first the adjacency matrix is randomly generated and then the corresponding graph is displayed).

A *help* function provides hints for the solution of the problem and the explanation of the manipulation techniques. If the problem is a difficult one, the *help* function can ask to solve easier subproblems while tracing the solution of the given one.

4.2 Equipment

The system has been implemented in Pascal on a personal computer Olivetti M24 (IBM PC compatible) under the operating system MS-DOS 3.1. The required hardware configuration is a computer with two disk drives or one disk drive and one hard disk.

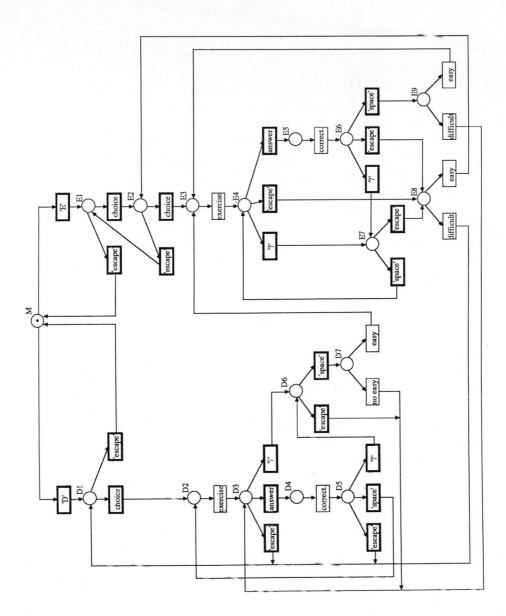

FIGURE 1

Petri net describing the system behaviour: the relation between difficult and easy exercises is shown in detail.

5. DESCRIPTION OF THE PETRI NET

In this net we have represented with heavy-lines boxes (transitions) the actions to be performed by the user and with light-lines boxes the actions performed by the system. The box labelled *exercise* corresponds to the generation of the exercise, while the *correct.* box denotes the comparison of the solution with the answer given by the user. Boxes having labels between apices denote the corresponding character entered by the user.

Circles (places) denote the different states of the system. The place labelled M corresponds to the initial state of system and to the visualization of the first menu that allows the user to choose between introductory, easy and difficult exercises (in the net the introductory module is omitted).

At place M transition $"D"$ and $"E"$ are enabled. The firing of transition $"D"$ leads to state D_1 where the menu of difficult exercises is displayed. The user may then choose the subject and the system reaches state D_2; that activates the generation of the problem. In the case of easy exercises the generation of the problem is activated after going through two different levels of menus (states E_1 and E_2).

At place D_7 the system fires the transition labelled *easy* if there exist easy exercises on the subject, and the transition *no easy* in the other case. In the same way, when at places E_8 and E_9 the transitions labelled *easy* or *difficult* are fired according to whether the system reached those states from an easy problem or from the explanation of a difficult one respectively.

ACKNOWLEDGEMENTS

The general lines of the GALOIS system follows those of the algebra course given at University of Milan by A.Bertoni in the academic years from 1981/82 up to 1984/85. We would like to thank dr U.Moscato of the Department for reading through the introduction to Logic, and all the students who helped with the implementation during their thesis works [10]: Mezzanotte, Morandi, Pasero, Tognotti, Vertuani.

REFERENCES

[1]   M.A.Arbib, A.J.Kfoury, R.N.Moll, *A Basis for Theoretical Computer Science* (Springer-Verlag, NewYork Heidelberg Berlin 1981).

[2]   A.Bertoni, G.Mauri, N.Sabadini, *Dispense di Algebra* (CLUED, Milano 1986).

[3]   N.L.Biggs, *Discrete Mathematics* (Clarendon Press, Oxford 1985).

[4]   C.L.Chang, R.C.T.Lee, *Symbolic Logic and Mechanical Theorem Proving* (Academic Press, New York 1973).

[5]   J.A.Goguen, J.Thatcher, E.Wagner, An Initial Algebra Approach to the Specification, Correctness and Implementation of Abstract Data Types, in: R.T.Yeh (ed.) *Current trends in programming methodology, IV: Data Structuring* (Prentice Hall, New Jersey 1978) 80-149.

[6]   R.D.Jenks, B.M.Trager, A Language for Computational Algebra, *ACM SIGPLAN Notices* **16**, 11 (1981) 22-29.

[7]   J.D. Lipson, *Elements of Algebra and Algebraic Computing* (Addison-Wesley, Readings 1981).

[8]   B.H. Liskov, S.N.Zilles, Specification techniques for data abstractions, *IEEE Trans. on Software Engineering* **1**(1975) 7-18.

[9]   W. Reisig, *Petri nets: an introduction,* EATCS Monographs in Theoretical Computer Science (Springer-Verlag, New York Heidelberg Berlin 1985).

[10]   T.Tognotti, C.Mezzanotte, A.M.Vertuani, A.Morandi, A.Pasero *Tesi di laurea in Scienze dell'Informazione,* Università degli Studi di Milano, 1984/85, 1985/86,1986/87.

[11]   Wand, M., *Induction, Recursion, and Programming* (North Holland, New York 1980).

ECM/87 - Educational Computing in Mathematics
T.F. Banchoff et al. (editors)
© Elsevier Science Publishers B.V. (North-Holland), 1988

QUALITATIVE STUDY OF DIFFERENTIAL EQUATIONS :   RESULTS FROM SOME
EXPERIMENTS WITH MICROCOMPUTERS

M. ARTIGUE, UNIVERSITE PARIS 7, IREM

V. GAUTHERON, UNIVERSITE PARIS 7

P. SENTENAC, UNIVERSITE PARIS 11

Abstract: In France, teaching about differential equations for
undergraduate students has not been influenced by mathematical and
technological developments. As a result, this teaching is, at the
present time, obsolete. This paper is concerned in its renewal,
specially in the viability of a qualitative approach at this
level. We present three courses, organized in   french
universities, and we try to evaluate the impact of the use of
computers in these courses.

1 : INTRODUCTION

1.1 : THE EVOLUTION OF THE SCIENTIFIC FIELD

Since the 17th century, the theory of differential equations has evolved
in several settings :

- the *algebraic* setting (finding exact algebraic formulae   for
solutions),

- the *numerical* setting (finding numerical   approximations for
solutions),

- the *qualitative* setting (global study of the properties of the flow).

The   algebraic approach has predominated during   most   of the development
of the theory, while the qualitative approach first evolved with H. Poincaré
at the beginning of the 20th century. However, it is clear that the theory
of dynamical systems in mathematics (as initiated by H. Poincaré) and the
computational power of   modern computors   have drastically changed the
subject matter during the last twenty years.

1.2 : THE INERTIA OF TEACHING

Until recently, the first years of university teaching have been little
influenced by this change, at least in France. Teaching still has as its
main object the finding of explicit solutions of differential equations.
Most instructional time is devoted to integrating equations in classical
cases (when such an integration is possible), and other equations are treated
by developing solutions into power series or Fourier series.

A   change   is now taking place.   New curricula for   the   *Classes*

Preparatoires aux Grandes Ecoles (1984) make explicit mention of qualitative and numerical studies. In some universities, during their introduction to computer science in the first two years, students have to write programs for numerical solutions of differential equations. But usually the activities which take place in this framework are unrelated to what is taught in the classroom. So, the use of the computer as a teaching tool is ineffective.

In effect, the didactical problems of adapting teaching to the scientific and technological evolution in this field, are still to be solved :

- How does one construct instructional sequences properly relating the different settings : algebraic, numerical, qualitative ?

- Is this possible at every level of instruction ? Under which conditions ?

- What are the obstacles to the renovation of obsolete teaching practices ?

- What might be the role of computers ?

Three pilot courses, organized in 1986-1987 in the universities of Lille I, Paris 7 and Paris 11 (Orsay) have been analysed. In what follows, we shalltry to obtain from this analysis partial answers to the above questions, especially the last one.

2 : PRESENTATION OF THE THREE PROGRAMS

The three programs presented here were organized with some common goals :

- To widen instruction in this area to other settings than the traditional algebraic one and in particular to the qualitative setting, by organising interplays between settings.

- To take advantage of the computer to encourage an experimental approach in the qualitative and numerical settings.

However, they differ in several features : the level of the students involved, the time available for instruction, the computational equipment available and its possible management, not to mention other aspects.
Each course, with its specific features, is outlined below.

2.1 : LILLE I

Level : first year of university (DEUG I).

Teaching time : 4 weeks i.e. 32 hours (10 hours of lectures, 20 hours of exercises and only 2 hours with micro-computers).

Number of students : 90

students divided into three groupsfor exercises

 .*Curriculum* : modelling situations through differential equations ;
integration of equations with separated variables and first order linear
equations ; introduction to the qualitative approach ; Euler method.
This class was part of the program of study in mathematics during the fall
semester.

Use of computers : mostly postponed : the students work on drawings
of fields or flows provided by the teacher. The one session of interactive
work with microcomputer was devoted to qualitative exploration (software
provided).

2.2 : PARIS 7

Level : third year of university ("licence" in mathematics) ; most of
these students had some previous experience withcomputers (50 hours at least,
often in BASIC).

Teaching time : 14 weeks, each week including 4 hours of lectures, 4
hours of exercises , 2 hours with microcomputers and free access to
computers.

Number of students : 60 (4 groups for exercises and computing).

Curriculum : Introduction to Pascal language (One weekly lecture of two
hours for 5 weeks) ; Differential equations : Existence and uniqueness
theorem (proved by convergence of Euler's method) ; integration of some
classical integrable cases ; qualitative study (isoclines, fences,
comparison theorems) ; numerical study (classical methods, order, error
estimate, variable step methods) ; differential systems : linear systems,
autonomous systems,stability of equilibrium points.

Use of computers : Interactive use : At the beginning, students were
provided with Turbo-Pascal, with graphics tools and software for drawing
vector fields and solutions of differential equations and autonomous
systems ; later, they had to write on their own small Turbo-Pascal programs
in order to tackle various questions : comparative study of numerical
methods , location of separatrices, life-time estimates ...

The students devoted the last four sessions with computers to a
project: usually the study of a specific equation, with or without a
parameter; the study may be of numerical type (estimate of life-time,
location of zeros or separatrices, location of limit cycles and computation
of their period..), or of graphical type (qualitative study of the
dependence on a parameter). These projects were intended to provide an
opportunity to integrate anexperimental approach with theoretical study.

2.3 : PARIS 11

Level   :   third   year   of   university   ("licence" in   applied mathematics).

Teaching time : 14 weeks, each week including 3 hours of lectures, 4 hours of exercises and 1 hour with computers ; free access to computers possible.

Number of students : 60, in 2 groups for exercises and computing.

Curriculum   :   This was a course in calculus, including   topology, differential   calculus   from R^n to R^p and   study   of   autonomous vector fields in R^n : theorem of existence and uniqueness, continuity with respect to   initial   conditions,   linear   systems   of   order   2,   linearisation, perturbations (saddle points and nodes), life-time of solutions. Seven weeks were devoted to the study of vector fields.

Use of computers : Most students had previous knowledge of Pascal, the others attended a one week session at the beginning of the year.

Students were provided with Turbo Pascal and a graphics library but no other software.   They first had to write a program allowing the graphical study of sequences defined by induction ($x_{n+1} = x_n + h$, $y_{n+1} = y_n + hF(x_n,y_n)$) ; this program was used later for the qualitative study of autonomous systems in R^2 (and improved when needed).

3 : COMPUTERS AND TEACHING

Due   to   external constraints, the use of computers in   these three courses was of three different modes :

    -postponed use,

    -interactive use with software provided,

    -interactive use with software written by students.

What were the advantages of each ? What did each achieve ?More generally, what hypothesis can be made on the basis of these experiments concerning the role, the costs and the limits of each ?

3.1 : POSTPONED USE MODE

This was   the   case in Lille because of   strict   constraints concerning the equipment.

This mode revealed itself to be well adapted to   various situations, in particular to the introduction of the qualitative approach.

The pedagogical tool used for this introduction was *matching* : students were provided with a list of equations and a set of drawings of flows

or   direction fields.   They   had   to   match drawings with equations and
justify their choice.   For   instance, for   the   second session of exercises
in Lille,   students had   to match drawings of flows (two of them are shown
below) with the following equations :

$$y'=y/(x+1)(x-1) \qquad y'=y^2-1 \qquad y'=2x+y \qquad y'=\sin(xy)$$
$$y'=\sin(3x)/(1-x^2) \qquad y'=(\sin x)(\sin y) \qquad y'=y+1$$

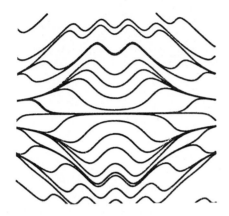 

Note   that   this choice is the result of a   didactical   engineery intended to
optimize the functioning of this type of situation in the given context [4].
For   instance,   there   are 8 drawings :   this number effects   a compromise
between various constraints :   making the matching   a real   problem,   giving
a wide range of drawings in order to   favour the use of various criteria,
and,   at the same time,   allowing students to obtain a solution within   a
reasonable time.   The characteristics of the drawings   are chosen so as to
make   matching by mere   gestalt   analogy difficult,   to require   actual
mathematical discussion and to favour interplay between the algebraic and
qualitative   settings.

   The complexity allowed by the computer plays a psychologicalrole as well :
 beyond simple esthetic interest,   the mastering of an apparently complex
situation is rewarding.

   The   experiment confirmed the hypothesis made regarding the   functioning
of this situation and its efficiency with beginners.

   Postponed use also seems to be a good tool in tasks of interpretation and
justification of drawings too.   In this context, one may imagine various
activities, for instance :
   - provided   with a   drawing,   students are asked to determine the
different types of solutions,   specify their   characteristics   and   justify
the interpretation given (this situation has been used for evaluation in

Paris 11),

    - provided with a drawing, students are asked to list questions for which the drawing does not give any answer and to conduct a theoretical study,

    - students are asked to complete a drawing in a justified way (note that separatrices often do not appear on computer drawings).

    - students are asked to make a comparative analysis of *neighbouring* equations.

We stress that for these tasks, postponed use of the computer has some advantages over interactive use :

    - it requires almost no equipment,

    - it allows a traditional managment of instructional sequences,

    - above all, one may focus the activity of students on the mathematical task by disassociating it from communication with the machine.

3.2: INTERACTIVE USE MODE

The postponed use mode has some advantages, as shown abovebut it has limits too : for instance it does not allow for an experimental approach towards phenomena based on exploration.

During the experiment, one type of situation turned out to be very fruitful and well adapted for this purpose : the qualitative study of differential systems depending on a parameter.

This study was undertaken both in Paris 7, in the framework of projects, and in Paris 11, in sessions with computers whose aim was to prepare for the study of linear systems and linearisation theorems.

In the latter case, instruction was divided into two phases :

    1) the study of the orbits of the autonomous system :

        x'=2x+my   y'=x+(2+m/2)y   according to the values of m

    2) the study of systems obtained from the previous one by perturbing it, adding terms of higher order (each student had to choose his/her own perturbation).

For the given system, the critical values of "m" are : -16, -8 and 0 and the following succession summarizes the evolution of the phase portrait:

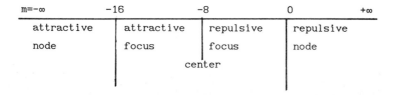

m=-∞	-16	-8	0	+∞
attractive node	attractive focus	repulsive focus	repulsive node	
		center		

By perturbations of higher order, non degenerate nodes and foci are preserved. This is not at all general for centers ; for degenerate nodes (m=-16,m=0), the situation is more complex.

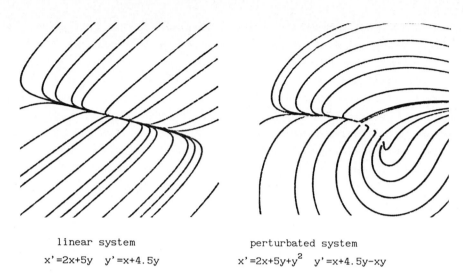

linear system   perturbated system
x'=2x+5y   y'=x+4.5y   x'=2x+5y+y^2 y'=x+4.5y-xy

Experimentation has confirmed the expected interest of this situation. The fact that some of the critical values were out of the usual domain reaffirmed the necessity of a theoretical approach to limit tedious random explorations. And the freedom of the choice for the perturbation in phase 2 provided straightaway a large range of situations, allowing a fruitful comparative study in order to point out invariants.

However, we must stress that even if the theoretical study is completed quickly and correctly, problems of framing and the choice of *good* values for the parameter (i.e. values leading to a fine catalogue of readable drawings), are time consuming.

Hence, this kind of activity will result in management problems, even with software provided, if students can't have free access to computers. On the other hand, it adapts itself to projects very well.

4 : EVALUATION AND CONCLUSION

Some data allow us to evaluate the impact of the use of computers in these three courses : observation of sessions, papers of students, examination papers...

At first glance, some features are striking :

4.1: CONCERNING THE QUALITATIVE APPROACH.

The use of computers seems to be crucial here, for several reasons :

    -it really gives meaning to the problematic of the qualitative approach,

    -it makes this approach accessible at an elementary level.

Indeed, the study of a differential equation *by hand* requires drawing isoclines, decomposing the plane according to the sign of Students poor ability to graphs and such things is a serious restriction on the viability of the qualitative approach (drawing curves given by equations is in any case a difficult problem, for which no good algorithm is known as soon as the equations become a bit complicated).

Through the use of computers, as described above, one may disassociate the different phases of this study by providing tangent fields, isoclines, incomplete or complete drawings...and this really enhances the accessibility of the qualitative approach, especially with beginners

    - it may also help teachers not to distort the mathematical practice in this domain, when teaching.

It is a well known fact that the constraints of teaching necessarily change mathematical objects and practices : knowledge is broken into elementary blocks, which are taught according to a precise order, special attention is paid to algorithms... As a consequence, examination questions are structured as a succession of elementary problems the students solve following the given hints, without any autonomy.

The use of computers can help teachers to elaborate activities, like those mentioned above, related to more global problems and, nevertheless, within the scope of students.

But one must not conclude that the use of computers makes learning difficulties disappear.

For instance, in Lille, for the final test, the students were given the following equation :

$$y' = (1/1+x^2)^2 - y^2 .$$

 - 84% among them, correctly decomposed the plan according to the sign of y',
 - 73% succeeded in drawing two solutions at least (three were asked for given initial values).

But the success rate collapsed when proofs were required (at the most 10%).

The results of the final test in Paris 11 were, in some ways, similar : all students succeeded in matching systems with drawings of flows. Then they were asked to choose a system and determine the different types of solutions. Most of students worked out a precise study of the equilibrium points, using linearisation techniques but quite few produced a correct classication of the solutions of the system or a correct study of their asymptotic behaviour

It is clear that elementary qualitative study is difficult to teach : What is obvious ? What needs a proof ? What kind of proof ? Which notions and theorems are to be stressed ? There is no traditional canon to which we could refer. These difficulties are aggravated by the fact that: at the present time, at least in France, students and teachers are quite unfamiliar with qualitative reasoning.

One may think that computers make the students think that proofs are not necessary, that they are only a matter of fulfilling a "didactic contract" with teachers. This is not our opinion : pedagogical situations like those mentioned in 3.1, when carefully managed, can lead students to gain necessary confidence in drawingsand, at the same time, to detect ambiguities and form hypotheses ; such situations thus can favour reasoning

4.2 : CONCERNING THE NUMERICAL APPROACH.

This was specially considered in Paris 7. In this case the use of computers is necessary, but the results of the experiment are less conclusive aswe have not yet found activities that call for theoretical control and are both easy and motivating enough for beginners, as we did for the qualitative approach.

This is part of our planed research and we will develop next year specific software to support such activities.

ACKNOWLEDGEMENTS

We should like to thank Mr. Sacré, Mr. Duflo and Mr. Isambert who wrote the softwares provided to the students.

REFERENCES

[1] Artigue,M. and Gautheron,V. Systèmes Différentiels : Etude Graphique (Cedic, Paris, 1983).

[2] Hubbard,J.and West,B. : Ordinary differential equations (Springer Verlag, Berlin, in print).

[3] Kocak,H. : Differential and difference equations through computer experiments (Springer Verlag, Berlin, 1983).

[4] Artigue,M. : Didactical engineery for differential equations in Acts of Psychology of Mathematics Education XI, Montreal, 1987, in print.

ECM/87 - Educational Computing in Mathematics
T.F. Banchoff et al. (editors)
© Elsevier Science Publishers B.V. (North-Holland), 1988

THE CALM PROJECT(*)

C. E. BEEVERS, A. R. AGNEW, B. S. G. CHERRY, D. E. R. CLARK, M. G. FOSTER,
G. R. McGUIRE, J. H. RENSHAW

Department of Mathematics,
Heriot-Watt University,
Edinburgh, Scotland,
United Kingdom.

ABSTRACT

The CALM Project seeks to·produce computer enhanced packages which will
consolidate the conventional teaching of Calculus to large groups of
first year engineering undergraduates at a typical Scottish University.
This article will describe a number of software tools designed for the
preparation of Computer Aided Learning materials (CAL) in Mathematics.
It will explain the philosophy and the features of software design
favoured by the CALM Project team.   The CALM Project provides a
practical example of software development in the programming language
of Pascal in an educational environment.

1. INTRODUCTION

   The Heriot-Watt University in Edinburgh is a small technological university
with large classes of engineering and science students.   Its Mathematics
Department has a major service teaching commitment to both the science and
engineering faculties.   The size of the service groups causes teaching
problems particularly in tutorial classes.   The lack of adequate resources has
exacerbated the situation so that the current tutorial arrangements for first
year undergraduates has become ineffective.   It should be noted that a typical
Scottish degree is taken over four years and that the majority of students
have seen very little Calculus before they come to University.   So, in the
first year of their degree our groups of engineering and science   students
have six hours of Mathematics per week   with the time split equally between
Algebra and Calculus.

   The experience of those of us teaching the large service classes has been
that the organisation is at best inefficient and at worst ineffective.   Large
lecture classes combined with unwieldy traditional pen/paper tutorials fail to
motivate the bulk of students.   Too many of them opt out of the tutorial
system and they do not fulfil their true potential.

   A new approach was needed.   Some of us had noticed that when the computer
was introduced into the curriculum student motivation increased (see [1 - 3]
for details).   Then, in Britain, following the Nelson Report [4] in 1983   the

*The CALM Project is funded by the Computer Board of the UK as part of a
Computers in Teaching Initiative

Computers in Teaching Initiative was set up to provide the funding for a
large variety of teaching schemes across the whole university subject
spectrum.   We recognised an opportunity to try a radical remedy to our
problems with service Mathematics. So, we put forward the following proposal:
we would design a computerised tutorial system to replace the conventional
tutorials for the teaching of first-year Calculus.   Our proposal was accepted
and the money we received from both the UK Computer Board and our own
university had two purposes: it would buy both the hardware for the computer
tutorials and the expertise to write the software.

   This article describes in the next section the basic philosophy of the CALM
Project.   Then, in section 3, we will concentrate on some of the mathematical
aspects of our project.   This includes a description of the software tools we
have developed.   In section 4 we will summarise some of the lessons we have
learnt from the students who have used the CALM software in the academic year
1986/7.   In the last section we will look ahead to the tasks yet to be
completed.

2. THE CALM PHILOSOPHY

   We chose the Research Machines Nimbus which is a British machine.   We have
a networked laboratory of 32 micros served by a 16-megabyte file server.   Each
micro can stand alone and there is a mix of single and double disc drive
machines with each one having half a megabyte of memory.   We also have 2
external Winchesters attached to the file server to give a further
80-megabytes of storage.

   We envisaged that each unit of software prepared would have three
components: theory screens to back up the conventional lecture, a worked
examples section to show the students how to do typical Calculus problems and
a test section to monitor student progress.   Each week of a 25 week course
(covering the fundamentals of differentiation and integration and an
introduction to numerical analysis and differential equations) is packaged
into a unit of software.   The units may contain more than one topic but they
are all constructed to the formula: theory, worked examples and test.   There
may be a number of theory and worked example programs covering the topics
encountered in that week but there is just one weekly test comprising
questions from all the topics of that week.   We write short programs in
general so that the whole system is not set in concrete.   This ensures that
changes can be made in subsequent years without too much difficulty.

   There are seven people in the CALM Project team and all have a programming
and mathematical background though to different degrees.   We decided to work
with the language Pascal for a number of reasons.   Firstly, after talking to
others in the area of CAL before us it became clear that we needed to choose a

language that has portability across a range of micros and secondly it came free! As a bonus the structured nature of Pascal is appealing to mathematicians. We have worked as a team throughout and it soon became apparent that common procedures could be used by us all. At the beginning these common routines were pulled into each individual program at compilation time. Now that this library of common routines has settled down we have an object code version of them. This library is linked selectively into the individual programs to produce an executable file to run over the network.

3. MATHEMATICAL SOFTWARE FOR CAL

One of the first general routines written for the library was a menu design program. This allows access to the software through a series of attractively designed screens. Such applications are easy to use and popular with students. The menus use coloured text with layouts difficult to achieve in print. The advent of coloured graphics has added a new dimension to the presentation of mathematical material. A flexibility of expression which shows itself in three main ways is now available to the teacher. Firstly, whole lines of text or formulae can be highlighted in significant colours or modes and individual symbols within a formula can be picked out very simply by colour. Thus, the global logic of a mathematical argument can be indicated merely by the selection of appropriate colours. More specifically, the progress of a particularly significant symbol (e.g. sign, variable or function) can be labelled on the screen in an outstanding colour and then its exact location in a whole row of equations, can be followed with comparative ease.

Secondly, particularly important symbols or instructions may be highlighted dynamically by means of the intermittent flashing of the symbol or instruction in colour. This 'dynamic' technique represents a radical departure from what is possible in a book. At a more advanced level, a significant change of sign (as when like terms are grouped together on one side of an equation, or when a common factor is removed from a collection of terms) can be signalled 'dynamically' by a flashing '+' or '-' sign or flashing the common factor. Finally, by consistently colouring one particular variable with the same colour, its progress during the course of a relatively complex piece of mathematical reasoning can be followed easily.

Thirdly, at any point in the course of the presentation of a piece of mathematics pertinent comments can be interjected temporarily on the screen. Each comment may be contained in attractively shaped windows, with moving arrows leading the eye to the point of interest on the screen.

All of these techniques have been exploited within the CALM system. The dynamic features of microcomputers can be used in other ways too. For example we have used graphics to broaden the students' experience. Graphs of functions can be quickly sketched on the screen to give visual emphasis. Moving tangents have been employed to indicate stationary points and points of inflexion. Mathematical 'games' have been constructed with student interaction an important feature. This form of learning appears throughout the CAL literature and its use was encouraged in the Nelson Report [4]. It provides an example of how the computer may be employed to good effect in teaching. Students learn when they are enjoying the material so much of what we do is intended to stimulate the learning process. Sprites have been designed to add further visual stimulus.

To achieve all of this the software designers needed routines to help in the preparation of the theory and worked example sections. Programs were written to handle text, place it on the screen at different sizes, colours and position. We have a screen editor and some short procedures to do simple jobs like underlining.

We perceived at an early stage that an important requirement was a way of checking formula. We decided to avoid string recognition and chose instead to write an evaluation routine to find the value of a given string. Another answer string, differently arranged, could then be compared with the 'correct string' by evaluating the two strings over a suitable range. This procedure is the basis of the test section allowing students to input their answers and receive confirmation of these answers almost instantaneously. The evaluation procedure is recursive and is capable of handling many-variable functions.

Our evaluation procedure initially checks the input string for simple errors, such as mismatched brackets, and for obvious sources of error, such as very long input strings. If such mistakes are detected the user is asked to correct the error or simplify the answer. The procedure then evaluates the string using an array of previously-stored variables and a wide range of mathematical functions, allowing for different ways of inputting especially parts like multiplication of factors and powers.

The test section software is written to give the students the chance to test their knowledge of a specific set of topics on offer in a unit. The student can take one or more tests in the privacy of the CALM laboratory. Ideally, the tests are done after the student has attended lectures on the topics, worked through some examples at home and after a revision of the theory and worked examples available on the computer. Evidence gathered in the summer term confirms that when in the laboratory the majority of students begin by refreshing their memory with a look through the theory and worked example sections but then they spend 75% of their time doing tests. The test program is designed to provide a record of the students' progress through the course by filing away the answers to the tests, marking them and storing a mark for each test.   The files containing the answers and marks allow both the teacher and student to judge how the course is going.   It could provide a continuous assessment of student performance if we wish.

Currently two levels of test are on offer in each week's unit: they are referred to as the easy test and the hard test.   The test questions are selected randomly from libraries of files.   Each file contains one question consisting of three parts: the statement of the question, some answer preambles to prompt a student's response and the correct answers with associated information to allow the evaluation routine to operate.   A question may have many parts and hence require a number of answers.   The answer preambles are an important feature as they are used to guide the student to input the answer in a particular way.   The staging of questions is a vital ingredient of the learning process of prime significance in the easy test.

In setting up the rules for the tests the number of 'attempts' allowed for any answer in a test question can be defined.   We have found that 3 'attempts' during an easy test and 1 'attempt' for a hard test works well in practice. To simulate the process of a student crossing out an attempted answer in a written examination paper we allow the student to restart a question.   The number of restarts permitted for any question in a test can also be defined. It was found that 2 restarts per question in the easy test were reasonable and that the number of restarts per question in the hard test should be greater than three.   During a test a student can press the escape key to gain access to other facilities such as a calculator, a graphics window, a list of standard integrals and information pertaining to performance in the tests. When the student has the information sought the escape key again guides the student back to the point in the program at which the exit was originally made.

The random selection of questions means that normally each student gets a different set of questions to do and also any one student can be given many different tests in any test section.   Even with only twenty questions this can be achieved by building in a random feature within some questions.

When the student has input answers to all the parts of the question, the true answers are displayed so that they can be compared with their answers. Finally, the set of answers is marked against the correct ones and the score for that question is shown.   The student's answers and the correct answers to the question are then filed away on the network.   The mark recorded in the test is also stored.

Clearly the marking and the filing away of the students' responses to the test questions would be useless if we could not examine these marks at a later date.   To this end we have developed a program, called mark, which consists of a number of procedures to allow us to 'see' the students' attempts and marks and so determine how well they are progressing. On entering the mark program the teacher is prompted with the units on view and invited to choose one of these units.   Details on the whole class or individual students can be displayed at the press of a key. Further details on all these software tools can be found in [5].

As well as the software tools detailed above we now have units of software available on the following topics: 1. Algebraic Limits, 2. Trigonometric Limits and Discontinuous Functions, 3. The Rules of Differentiation, 4. Inverse and Implicit Functions, 5. Tangents/normals and related rates, 6. The Theory of Max/min, 7. Practical Max/min Problems, 8. Series, Hyperbolic Functions and Harder Differentiation, 9. Parametric Differentiation and L'Hopital's Rule, 21. Numerical Integration, 22. Numerical solution of equations, 23. First Order ODEs of Variables Separable and Homogeneous Type, 24. First Order Linear ODEs, Bernoulli's Equation and 2nd order reducible equations.   Units 11 - 19 covering the elements of integration are well underway but they have not had a thorough testing yet. In addition, units 10, 20 and 25 will be revision packs on differentiation, integration and ordinary differential equations respectively.   Thus, each of the 25 units covers a normal week's work over the 25 week duration of our elementary introduction to Calculus.

4. STUDENT EVALUATION

We have listened to the students throughout and we embarked on a formative evaluation of the CALM software.   The detailed description of this evaluation appear elsewhere [6] but here is a summary. Through a series of interviews and questionnaires a number of worthwhile suggestions emerged. For example, they prefer 80-character width screens in general and white text on a black background for the bulk of screen information. This is not to deny either the use of colour or larger-sized characters. On the contrary we have used both effectively to highlight specific mathematical points.

The students reported that at the start they were apprehensive about using the computer but after nine weeks use this worry had largely disappeared. One of the main reasons for this change appears to be the success of the computer in overcoming their embarrassment.   They told us that they felt happier displaying their mistakes to the machine than to the tutor.   They looked forward to the tutorials and enjoyed the weekly challenge embodied in the tests.

The attendance in the tutorials seemed to bear these facts out since we maintained a 90% turnout throughout the term whereas conventional tutorials running in parallel could only manage a 50% attendance.   The quiet, concentrated work in the computerised tutorial indicated a more effective learning environment. Much of this has been confirmed by an independent report by McClemont [7].

5. THE CALM FUTURE

We intend to add to the on-line information to include further graphical capabilities. We also want to experiment with the use of help or hint screens to prompt a student along the right road of solution in a worked example. There are other parts of the course yet to cover but we expect to complete our original 3-year project by October 1988 exactly to schedule.

ACKNOWLEDGMENT

The authors are grateful to the Computer Board of the UK and the Heriot-Watt University for funding our project.

REFERENCES

[1] Beevers, C. E., Motivating mechanics, IMA J. of Math Teach., 4 (1985) 52.
[2] Beevers, C. E., Mechanics with a micro, Proc 2nd Int. Conf on Math Modelling, Exeter University, (D. Burghese, 1985) 217-223.
[3] Beevers, C. E. and Clark, D. E. R., The micro-revolution in mathematical education, Proc. Workshop on Computer Based Teaching, Cambridge University (1985).
[4] Nelson, A., The Nelson Report, (1983).
[5] Beevers, C. E. et al, Software Tools for Computer Aided Learning in Mathematics, Int. J. Math. Educ. Sci. Tech, in print.
[6] Beevers, C. E. et al, The CALM Before the Storm! Computers and Education 12 (1987) in print.
[7] McClemont, S, CALM - an educational evaluation, Department of Civil Engineering, Heriot-Watt University, (1987).

ECM/87 - Educational Computing in Mathematics
T.F. Banchoff et al. (editors)
© Elsevier Science Publishers B.V. (North-Holland), 1988

EXPERIMENTAL COMPUTATIONS: A MAIEUTIC APPROACH TO ABSTRACT MATHEMATICS

Paolo BOIERI

Dipartimento di Matematica - Politecnico di Torino
Corso Duca degli Abruzzi, 24
10129 Torino, Italy

The traditional methods of teaching Mathematics to applied science students are considered, pointing out that they do not provide a correct balance between theory and examples. The use of computers, by presenting an adequate number of new and stimulating situations, can motivate the exploration of a wider field of mathematical theory. Two examples of the use of computers are presented, based on the author's teaching experience.

1. INTRODUCTION

In this paper, an experience of using computers in teaching Mathematics to Engineering and other applied science students is presented, emphasizing the underlying philosophy that induced some teachers to introduce and use computers in their courses at the Polytechnic of Turin.

The traditional method of teaching Mathematics to undergraduate Engineering students is considered, pointing out that it is usually based on two different approaches.

In the first approach, that, roughly speaking, could be called *application oriented*, the mathematical problems are introduced by means of a series of elementary examples suggested by applications to Physics, Biology, Economics, etc. The mathematical theory is presented bearing in mind the practical purposes of the training.

In the second approach, that could be called *theory oriented*, the mathematical theories are treated rigorously, starting from algebraic and topological axioms, with limited relation to applications.

A careful analysis of both approaches shows that they have some relevant disadvantages; in particular, they do not offer a balance between the theory presented and the problems where this theory is applied. These problems are usually limited in number and in quality (and too closely related to applications in the first approach); they tend to confine the student to the use of fixed recipes, without a thorough understanding of the subject.

In this context, the use of computers helps the teacher in introducing new **facts and situations that cannot be** immediately reduced to already known cate-

gories. Such thought provoking situations result in the need for new theoretical answers, that motivate, in turn, new exercises. In this way the use of
computers is not an alternative to theory; on the contrary, it is a substantial help in enhancing the quality and quantity of Mathematics taught.

These possibilities of using computers are illustrated by two examples.

2. WHY AND HOW COMPUTERS SHOULD BE USED

2.1. Two approaches to teaching Mathematics

The main features of the *application oriented* approach can be summarized as
follows: a certain context of problems is presented, originating from Physics,
Biology, Economics or other applied fields, with many details.

These problems can be solved within a specific mathematical area: for
instance, the dynamics of populations requires nonlinear ordinary differential
equations and the propagation of electromagnetic waves requires a good
knowledge of tensor calculus.

These mathematical topics are presented to students, explaining only what
is relative to each application, without involving the theoretical foundations. A couple of examples may be useful to clarify this point. The theorem
of existence and uniqueness of solutions is fundamental in the study of ordinary differential equations. However, in a course in biological modeling it is
only stated (but not proved) and usually not used, since the applications
seldom involve very critical situations. In a course in electromagnetic wave
propagation tensor calculus is necessary, but its foundation on differential
geometry is not presented to Engineering students.

All the mathematical theory presented is immediately applied to the
solution of problems suggested by applications and, at the end of the course,
the student is required, in a written examination, to solve some of them.

A schematic diagram of this way of teaching can be the following, that
shows a "closed loop" structure: applications, mathematical theory, problems
connected with applications.

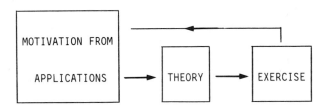

FIGURE 1

This approach has many merits: first of all, it is very effective, since a considerable amount of mathematical knowledge is taught in a relatively short time. For this reason and for the insufficient time devoted to Mathematics in Engineering courses, it is usually a necessary choice.

Moreover, the close connection with applications is very useful to the student, who is easily motivated in the study of Mathematics.

But some relevant disadvantages must also be pointed out. Since the theory has been developed in view of a particular field, it is very difficult for the student to face new situations that may arise in applications. He can either misuse the mathematical tools with which he is familiar, i.e. use them in domains where they are not valid, or not understand that the new application he is considering requires such tools.

The final result of this teaching method runs the risk of being only a collection of recipes for the solution of strictly determined problems. A high degree of efficiency and a certain amount of sterility are mixed in the *application oriented* approach.

The opposite attitude towards teaching Mathematics to applied science students is the one previously called the *theory oriented* approach.

Here the main role is played by the mathematical theory and not by its applications. Calculus may be chosen as an example. In this approach, calculus is developed starting from set theory and the algebraic and topological axioms of real numbers. The role of the axiom of completeness is particularly emphasized and the whole presentation is based on a sequence of theorems on local and global properties of functions, on series, integrals and diffe- rential equations.

All the theorems (or, at least, the majority of them) are proved and the oral part of examinations is important in the final judgment of the student's capabilities. In this context, the applications are not mentioned at all or they are treated very briefly.

This approach also has merits and drawbacks.

Among the merits, the first to be pointed out is that this teaching method is a very good exercise in rigor and logic and, regardless of the practical uses of the Mathematics taught, it helps the student's intellectual growth.

Since the domain of validity of the results is clearly stated and the critical situations carefully considered, the future user of Mathematics has a precise idea of the type of procedure -whether rigorous or intuitive- he is using in his applied field.

Further developments of the mathematical theory can be easily pursued, relying on the student's sound background, even in advanced fields such as functional analysis, partial differential equations, information theory, etc.

But here the teacher is facing a difficult practical problem: this approach is extremely time consuming. It is also difficult to justify the choice of axioms, the sequence of deductions and the sophistication of material presented to many Engineering students.

Moreover, there exists a more serious problem, connected with the theory-problem relation: usually the problems presented to the students are not sufficient, in quality and in quantity, to explore the wealth of the theory completely. They are limited to cases that can be solved by paper and pencil and they are not presented for their intrinsic interest but just as mere illustrations of the theory. In this way, they do not stimulate any further investigation.

In this situation, the main pedagogical requirement is then to re-balance the relationship between theory and examples, by giving the students more and more stimulating examples to support the theory.

2.2. The role of computers in mathematical courses

As we have seen, both approaches present many merits and some serious problems. Usually the way of teaching Mathematics to Engineering students is not a free choice for the Mathematics teacher, but it is strongly influenced by the general study curriculum, by the limitations in time and also by the cultural tradition of the country (for instance, in Europe, the *application oriented* approach is more diffused in the U.K. and Scandinavia, while the *theory oriented* approach is preferred in France and Italy).

But, regardless of the method chosen, the preceding discussion shows that the teacher willing to improve his work has to face an important problem: the student is not exposed to an adequate number of sufficiently stimulating problems and applications. This is due to the limitations of the "paper and pencil" method.

In this context, the appearance of computers in the educational world has changed the traditional approach completely.

The students who enter University generally already have some knowledge of computers: all of them have already played videogames, many can write simple Basic programs, some of them have used spreadsheets, databases or word processors.

Being interested in getting more examples for their mathematical courses, the students immediately try to use computers to produce them. But their approach to "Mathematics with computers" is usually naïve, as they are not able to combine mathematical with programming capabilities and usually have no knowledge of the fundamental concepts of numerical computations.

To avoid this misuse of computers in Mathematics, that can compromise a proper understanding of the theory involved, it is necessary to use computers in correctly supervised mathematical courses.

Computers can be used as "improved blackboards": they offer computing capabilities that enable the teacher to overcome the limitations of the "paper and pencil" problems, graphic capabilities that visualize many situations, otherwise difficult to understand, interactive capabilities that improve the effectiveness of books as learning aids.

But the most important point of this discussion is the following. The correct and creative use of computers is not only an effective help to teaching but is also a source for new and more advanced theories and stimulates the student to tackle new problems. The machines can offer students a critical mass of examples. On this basis some questions arise together with the necessity of new theoretical answers, that only a more advanced treatment of Mathematics can supply.

Thus the exercise is no longer a mere and sterile use of the theory or a passive application to non mathematical fields, but a source of problems that motivate the deepening of the mathematical theory: it has, in this way, a *maieutic nature*.

The diagram in Figure 2 shows this situation.

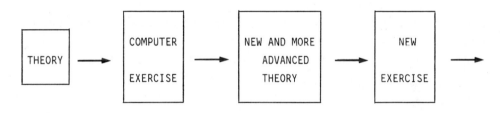

FIGURE 2

The above structure is no longer closed, as in Figure 1: it is an open structure, suggesting an infinite progression.

3. SOME EXAMPLES

Two examples are proposed now, arising from a calculus course. They refer to different situations: one is elementary, while the other is more advanced.

First example: the expansion of a rational number in a new base.

Let us consider a rational number $a = p/q$, with $0 < a < 1$ and a natural number $b > 1$. We want to write the representation of a as an expansion in base b.

This problem is important for the general comprehension of the concept of number and also for its applications to computer science, in the case $b = 2$.

It is solved by an algorithm: if a is written as

$$a = 0.d_{-1}d_{-2}d_{-3} \cdots,$$

the digits (in base b)

$$d_{-1},\ d_{-2},\ d_{-3},\ \ldots$$

are determined by the relations

$$p\quad b = d_{-1}\ q + s_{-1}$$
$$s_{-1}\ b = d_{-2}\ q + s_{-2}$$
$$s_{-2}\ b = d_{-3}\ q + s_{-3}$$
$$\ldots\ldots\ldots\qquad\qquad ,$$

with $0 \leqslant d_{-i} < b$ and $0 \leqslant s_{-i} < q$.

These relations can be used to solve our problem with paper and pencil. But you can hardly find a student who does not get bored after three or four applications and moreover some cases may occur that defy the most diligent students. On the contrary, it is easy to write a program that enables us to find an adequate number of interesting examples in a short time: some of them are collected in Figure 3.

```
Expansion of a=p/q, with 0<a<1, in a base b, 2≤b≤16.

Choose b, with  2≤b≤16          Choose b, with  2≤b≤16
? 10                            ? 2
How many digits?                How many digits?
? 10                            ? 20
Introduce p and q, with p<q     Introduce p and q, with p<q
? 1,10                          ? 1,10
In base 10 the expansion is:    In base 2 the expansion is:
1/10 = 0.1                      1/10 = 0.00011001100110011001
```

FIGURE 3

Looking at the computer output, some questions arise immediately:
a) Is it possible to predict when the expansion is finite and when it is infinite? In the second case, is it possible to predict the length of the period, without an actual computation of the expansion ?
b) How can this prediction be done computationally ?

The first question involves new and more sophisticated theory, while the second question motivates a computer exercise, more advanced than the one proposed above.

Second example: structural stability of linear systems of ODE's.
We consider a system of ODE's with constant coefficients

(3.1)
$$\begin{cases} x' = a\,x + b\,y \\ y' = c\,x + d\,y \end{cases}$$

where $x(t)$ and $y(t)$ are the unknown functions and a,b,c,d are real numbers. We want to study the effects of *small perturbations of coefficients* on the *asymptotic stability* properties of the rest points of the system.

The rest points are determined by solving the linear algebraic system

$$\begin{cases} a\,x + b\,y = 0 \\ c\,x + d\,y = 0 \end{cases},$$

corresponding to the matrix

(3.2)
$$A = \begin{pmatrix} a & b \\ c & d \end{pmatrix}.$$

Since only the non degenerate case (i.e. $\det A \neq 0$) is considered here, this system has only one solution: the origin. It is well known that asymptotic stability of the differential system (3.1) is determined by the eigenvalues of A: it is asymptotically stable when all eigenvalues of (3.2) have a negative real part, unstable when all eigenvalues have a positive real part and partially stable (saddle configuration) when the eigenvalues are real with opposite sign. There is also a center configuration (eigenvalues with zero real part) which is stable but not asymptotically stable. Degenerate configurations can also be studied (see, for instance, [3] for details).

A planar linear system with constant coefficients is uniquely determined by assigning a 2x2 matrix. The space of these matrices, denoted by $M(2)$, can be identified with R^4 endowed with the euclidean norm
$$\| A \| = (a^2+b^2+c^2+d^2)^{\frac{1}{2}}.$$
A small perturbation of A is a small change of its coefficients that induces a slight modification of $\| A \|$.

We can now make some computer experiments. Consider first the matrix

$$E = \begin{pmatrix} -1 & 1 \\ 0 & -1 \end{pmatrix}.$$

P. Boieri

In Figure 4 the corresponding phase space is plotted, together with all the relevant informations about configuration (a non degenerate stable node), eigenvalues and eigenvectors (the matrix E is in Jordan form with a double eigenvalue and one eigenvector whose direction is shown in Figure).

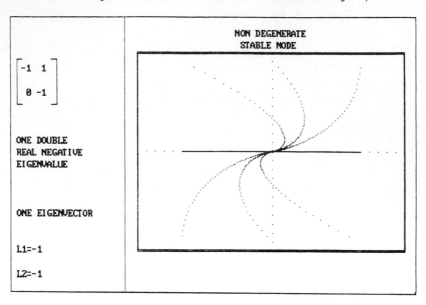

FIGURE 4

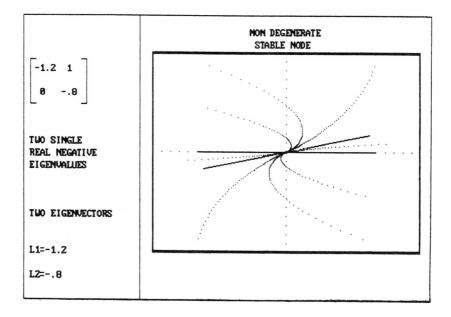

FIGURE 5

A slight change in the diagonal terms of E gives a new configuration (see Figure 5), with two eigenvalues and two eigenvectors, but the asymptotic stability is conserved.

Changing also the non diagonal terms, the focus configuration of Figure 6 is obtained. The eigenvalues are now complex and no real eigenvector exists, but the configuration is again asymptotically stable.

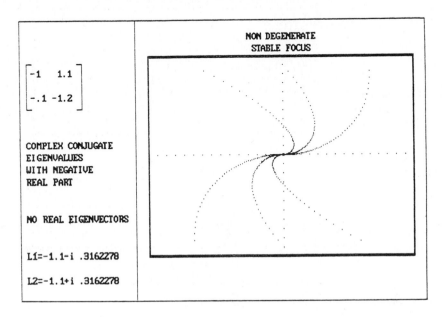

FIGURE 6

A further example can be considered; a center configuration is chosen now, corresponding to the matrix

$$F = \begin{pmatrix} 0 & -1 \\ 1 & 0 \end{pmatrix},$$

as shown in Figure 7.

Slightly perturbing the non diagonal terms, we have a focus configuration. But, in this case, stability is no longer conserved, since we can get either a stable focus (Figure 8) or an unstable one (Figure 9).

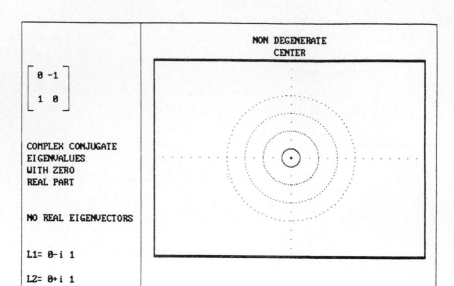

FIGURE 7

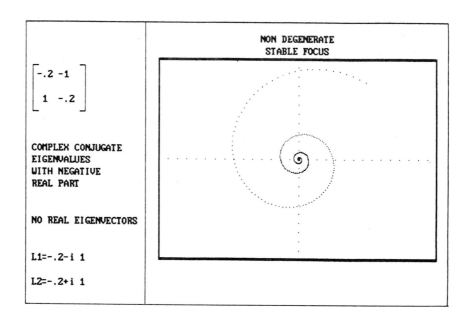

FIGURE 8

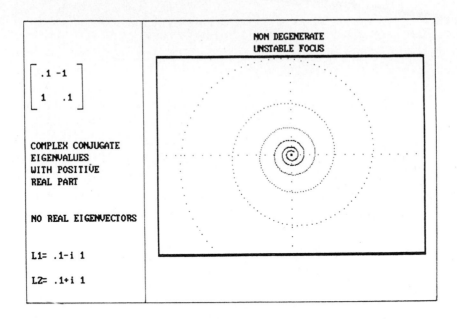

FIGURE 9

The computer experiment must be completed, considering unstable and saddle configurations. It can be noticed that, in these cases, the stability properties are conserved by small perturbations of the coefficients of the matrix, as it happens for the matrix E.

To explain the results obtained, we must introduce a new classification of the configurations, based on stability. Some fundamental concepts must be defined: the *linear* and *topological equivalence* of dynamical systems and their *structural stability*. We simply recall (see the references of [3] for a complete treatment of the subject) that the set M(2) can be divided in four subsets, according to the stability properties of the related systems. Each subset is invariant with respect to topological equivalence. The subsets corresponding to stable (resp. unstable and partially stable) configurations are open in M(2). On the contrary, the set of matrices corresponding to center configuration does not contain any open set: for this reason a small perturbation of the coefficients can produce a matrix that no longer belongs to the set.

It is evident from this example that a higher level theory is required in order to explain the results of the computer experiment.

4. FINAL REMARKS

The experience of using computers in Mathematics courses is still in progress at the Polytechnic of Turin. It involves first and second year students of Engineering with different curricula (Mechanics, Electronics, Civil Eng., etc.).

The high number of students in each course and the limited time do not allow to cover all the topics of the course using computer experiments: a selection is made following the interests of the students.

Students usually work at computers in small groups (usually, three students). They are required to write short programs (in Basic and Turbo Pascal, generally) and to discuss the mathematical questions arising from the experiment in a written report. The most important issues are then discussed by the teacher in class.

It may happen that particularly good students go further in their experiment work, writing nice programs, with careful error trapping and user interface routines.

But the principal aim is always the discussion of the mathematical problems and not the development of software. When the programming effort is too heavy compared with the mathematical benefits arising from the discussion, commercially available software is used or special packages are prepared by teachers.

The choice of the algorithm to be used is based on its theoretical relevance and a more careful analysis is usually left for higher level courses in numerical analysis, that students are encouraged to take.

The following references cover the most important publications of the group of teachers of the Polytechnic of Turin. In the references of the mentioned papers it is possible to find the works where inspiration for methodology and choice of the contents has been taken.

Many thanks are due to G.Geymonat for stimulating the author's interest in this subject and for many helpful discussions and suggestions. The author would also like to thank his colleagues at the Department of Mathematics of the Polytechnic of Turin, with whom he is sharing this experience: A. Bacciotti, L. Chiantini, M. Mascarello, L. Montrucchio, P. Moroni, A.R. Scarafiotti.

REFERENCES

[1] Bacciotti, A. and Boieri, P., Teaching Mathematics to Engineers: some remarks on the Italian case, in: Proceedings of the ICMI/ICSU-CTS Seminar,
Udine, 1987.

[2] Bacciotti, A., Boieri, P. and Moroni, P., An introduction to linear differential systems with the aid of a personal computer, Int.J.Math.Educ.
Sci.Technol. 17 (1986) 31.

[3] Bacciotti, A., Boieri, P. and Moroni, P., An introduction to structural
stability of linear differential systems, to appear in Int.J.Math.Educ.
Sci.Technol.

[4] Bacciotti, A., Boieri, P. and Moroni, P., A computer approach to nonlinear planar systems of ODE's, Comput. Educ. 11 (1987) 253

[5] Boieri, P., Some preliminary remarks to the use of computers in a Calculus course, to appear in: Proceedings of the Fourth European Seminar
on Mathematics in Engineering Education held in Goteborg, 1987.

[6] Geymonat, G., Congetture e dimostrazioni: possibili usi didattici dei
calcolatori tascabili nell'insegnamento della matematica, La Fisica nella
Scuola, XV (1982) 129.

[7] Geymonat, G., Lezioni di Matematica per allievi ingegneri.Vol.1 (Levrotto
e Bella, Torino, 1981).

[8] Mascarello, M., Winckelmann, B., Calculus and the computer. The interplay
of discrete numerical methods and calculus in the education of users of
Mathematics: considerations and experiences, in: Kahane, J.P. and Howson,
A.G., (eds) The influence of computers and informatics on Mathematics and
its Teaching, (Cambridge University Press, Cambridge, 1986) 120.

[9] Mascarello, M., Scarafiotti, A.R., Computer experiments on mathematical
analysis teaching at the Politecnico of Torino, in Supporting papers,ICMI,
(IREM, Strasbourg, 1985).

ECM/87 - Educational Computing in Mathematics
T.F. Banchoff et al. (editors)
© Elsevier Science Publishers B.V. (North-Holland), 1988

THE HOME MICROCOMPUTER AS AN AID TO DISTANCE TEACHING

T. M. Bromilow, J. A. Daniels, J. M. Greenberg

The Open University, Milton Keynes, England

The Open University's new Home Computing Policy comes into effect in 1988, when there will be a variety of courses involving the use of a microcomputer in the student's home. This paper describes how the microcomputer will be used in a course on Computational Mathematics.

1. BACKGROUND

The Open University is a distance teaching institution with 70,000 students currently studying part-time for a degree. Teaching is by correspondence text supported by television programmes/video cassettes and radio programmes/audio cassettes. For many years students have taken courses involving computing and doing their practical work using one of 300 remote terminals linked to central mainframe computers (see, for example, [1]). By the early 1980s, despite considerable enhancements to the system, it was clear that this method was unsatisfactory for four main reasons:

(i)   some students had to travel long distances to use a terminal;

(ii)   terminals were located in places such as libraries and teaching institutions so access was restricted;

(iii)   the breakdown of equipment caused frustration;

(iv)   as student demand for computing courses increased, a mainframe service could not support the required number of remote terminals.

The result has been a severe limitation on the amount of computing that the student could be expected to do. Two years ago the University decided that home computing using microcomputers was the only viable solution to the computing problem. After lengthy discussions with many manufacturers it was decided that no particular manufacturer could be given the contract for supplying microcomputers because of the volatility of the market. The following specification, satisfied by most IBM-PC compatible machines, was adopted to enable students to choose from a variety of machines:

MS DOS (operating system)

GEM (graphics environment manager)

512k RAM

Single disc drive

Keyboard (PC/AT)

Monochrome monitor

Parallel port (centronics: printer)

Serial port (RS232 or equivalent)

Mouse

Printer (80 column, 100 cps)

The use of a number of different machines to run course software has presented its own problems. The major differences in IBM-PC compatibles are in the use of graphics, since no accepted standard exists. We thus had to consider the graphics management systems commercially available which were supported on IBM compatible micros. The Graphics Environment Manager (GEM) from Digital Research was selected and a student environment was developed within the framework of the GEM Desktop software. This environment incorporates all the standard facilities within GEM Desktop, such as drop-down menus, windows, dialogue boxes, icons etc, using the mouse to select the appropriate menu item. All programs are written in Lattice C because of its extensive screen handling facilities and reliable bindings to the GEM software. Software is provided to students on floppy disks and they are expected to have installed the GEM Desktop software on their machines. The software will run on any IBM compatible micro which satisfies the Open University's Home Computing Policy specification. All the software development was carried out by the University's Academic Computing Service.

To minimize the difficulties which students may encounter working on their own, our objective was to develop portable, user-friendly and reliable software which would enhance the teaching of the course material.

2. M371: COMPUTATIONAL MATHEMATICS (first presentation: February 1988)

M371 [2] is an honours-level course which teaches those computational methods used in operational research. All the course material has been written and published by a small Course Team of academics at the Open University over a period of three years. The course content is presented in twelve correspondence texts (units) each requiring 12 - 15 hours of study time. The course is divided up into four blocks of three units as follows:

Block I: Solution of non-linear equations; systems of linear equations and systems of non-linear equations.

Block II: Linear and integer programming.

Block III: Non-linear optimization.

Block IV: Computer simulation of queues.

In all areas the objective is to teach the theory behind those methods which are currently used in commercial software so that, for example, the product form of the inverse basis method is described in the linear programming units while the DFP and BFGS methods are discussed in the non-linear optimization units. Case studies are also introduced to bring the subject to life and to put the theory into context.

The Course Team spent some considerable time discussing the most effective use of the home microcomputer to enrich the student's understanding of the course material. Two types of software were identified:

(i) **Teaching packages:** These would enhance the teaching of the more difficult parts of the course, making extensive use of graphics, and would be driven by an audio-cassette which would suggest lines of approach to a particular problem, explain some of the theory in an informal way or draw students' attention to important results which would be displayed on the monitor.

(ii) **Application packages:** These would enable the student to solve a variety of problems using sophisticated methods. The student would have the choice of exploring problems which are stored within the program or to input their own problems.

The two types of package are described in detail in the next two sections.

3. TEACHING PACKAGES

In comparing the 'teaching packages' described here with much of the educational software currently available, it should be borne in mind that our packages have been specifically designed to be studied by Open University students, unaided, in their own homes. In fact students will be studying *all* the materials for this course at their own pace, largely without tutor guidance or peer support, and may have to fit their study into their daily lives with frequent breaks or interruptions.

The teaching media used in M371: Computational Mathematics are:

Written materials – 12 units, handbook, computing booklet

Tutor contact   – correspondence tuition (four assignments)
                    face-to-face tuition (1 – 8 hours for the whole course)

Home computing facility
Audio-cassette tape

Application Packages

Teaching Packages

In order to appreciate how the teaching packages are used, it will help to consider first a teaching method which has been particularly successful in the distance learning of mathematics at the Open University: the combination of audio-cassette tapes with "tape frames" in printed text (see, for example, [3] and [4]). The student reads the printed text and is then instructed when to start the tape. While he/she listens to the tape, he/she looks at, works on and writes in the 'frames' drawn in the course text. The advantages of this method are:

1.   It helps to 'talk through' difficult techniques.

2.   The student is *involved*, completing the 'frames' his/herself.

3.   The frames can include graphs and diagrams, which may incorporate the student's answers.

4.   The student can study at his/her own pace.

5.   The student can rewind and replay the tape if in need of extra help.

6.   The student's work can be interrupted and readily resumed later.

In other words, the method's *flexible*. However there are two disadvantages:

1.   The tape frames are 'expensive' in pages of course text.

2.   Tape frames are *static*.

    The first idea in using the home micro with audio-cassette tape was to provide *moving* tape frames. But once we started to develop the idea, we realized that there was greater potential in it. Some of the advantages of using the home micro in this way are:

1.   Tape 'frames' can move.

2.   Many pages of printed text are saved.

3.   Calculations can be performed for the student.

4.   Full use can be made of the computer graphics.

5.   The keyboard can be used for student input, with instant response in the 'frame'.

    The student can now concentrate on the theory and the method instead of being distracted by getting involved in detailed calculations. More importantly, without tediously plotting graphs, he/she can see the graphical implications of what is being taught *using the student's own values* to experiment with. Values can be keyed in, in response to the spoken tape, and the mouse or keyboard can be used to control the pace.

    The role of the tape is both to explain the theory behind what the student sees, and to prompt the student on how to control the packages, via the mouse or keyboard. We minimise the amount of text on the screen so that the student can focus on the graphical and numerical effects.

    To demonstrate the ideas here we give examples from the first teaching package, on solving f(x) = 0. There are five 'screens' in this package: Bisection, Simple Iteration, Convergence Conditions and the Add nx Method, Newton-Raphson and Newton-Raphson viewed as Simple Iteration. The student would normally work through these in order but can, if required, 'quit' a screen and then return to the main menu, start a screen again or move on to the next screen. The options are made available in a 'drop-down menu' and are controlled by pointing and clicking the mouse. For example, for simple iteration the options are to 'input x_o', to start the iterations, or to 'quit'. Once the value for x_o has been provided, the student can choose any of the options at the top of the screen: 'next iteration', 'end iteration' or 'zoom' as shown in Figure 1. The main part of the screen is divided into two areas: the graphics are shown on the left and the text and numerical results on the right.

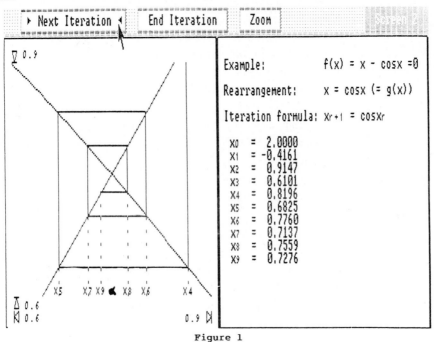

Figure 1

By clicking the mouse on 'next iteration', an iteration is performed, the calculated value shown on the right and the next step in the 'cobweb' diagram shown on the left. 'End iteration' will result in a message: "well done" if the required accuracy has been reached in a minimum number of iterations, how many iterations could have been used if the student has 'overshot', and so on. The complete cobweb diagram is shown when the student chooses to stop. 'Zoom' enlarges the relevant part of the graphics area to show the iterates in more detail. Thus at each stage the student is in control and can experiment and work at his/her own pace.

Note that we have *not* attempted to synchronise the tape with the progress of the software; when the student is required to listen to the tape, the screen shows a 'start tape' message, and when the student is required to stop listening to the tape and do something else with the computer, the tape will prompt him/her to choose an appropriate option.

The tape scripts were produced by BBC Open University Productions and close liaison is required between the authors in the course team, the software designers and the BBC producers for this teaching method to work and to produce it in a feasible timescale.

4. APPLICATION PACKAGES

The objectives of the application packages are

(i)   to encourage students to use the methods encountered in the units to solve a variety of problems;

(ii)   to allow students to gain insight into which methods work well and to understand why they sometimes fail;

(iii)   to explore the sensitivity of a problem to small changes in the data;

(iv)   to enable students to use commercially available software in a constructive and rational way.

To meet these objectives the application packages have been designed so that:

(i)   little computing expertise is required in order to be able to use them;

(ii)   great flexibility is allowed in the input or editing of a problem;

(iii)   results are provided which are easy to interpret;

(iv)   it is difficult, if not impossible, to make irredeemable errors;

(v)   good use is made of the graphics facilities offered by GEM.

Figures 2, 3 and 4 illustrate the screen layout for the application package for the first unit: finding roots of the equation $f(x) = 0$. The various menu titles are displayed along the top with the 'Options' menu shown in detail. The 'New problem' menu item will delete an existing problem and enable students to input a new problem (we also check that the student does not want to store an existing problem). The 'Tutor comments' menu item gives hints for the computing exercises set in the units.

The other menus are - 'File': to enable students to retrieve, store or delete problems; 'Help': to give advice if the student is stuck; 'Output': to control the amount of output which is displayed on the screen or is output to the printer.

When the student wants to input a new problem or edit an existing problem then the dialogue box appears, as in Figure 3, which allows students to choose the method they want to use. The method selected will determine what input is required to specify a particular problem. For example if the Newton-Raphson method is selected then the student must input both f and f' whereas the other methods do not require f'. Any information previously input, but not required by a particular method, will not be displayed but will be held in memory.

Copyright (C) 1987 The Open University.

Figure 2

▸ Options   ▸ File   ▸ Help   ▸ Output   AI_1

Figure 3

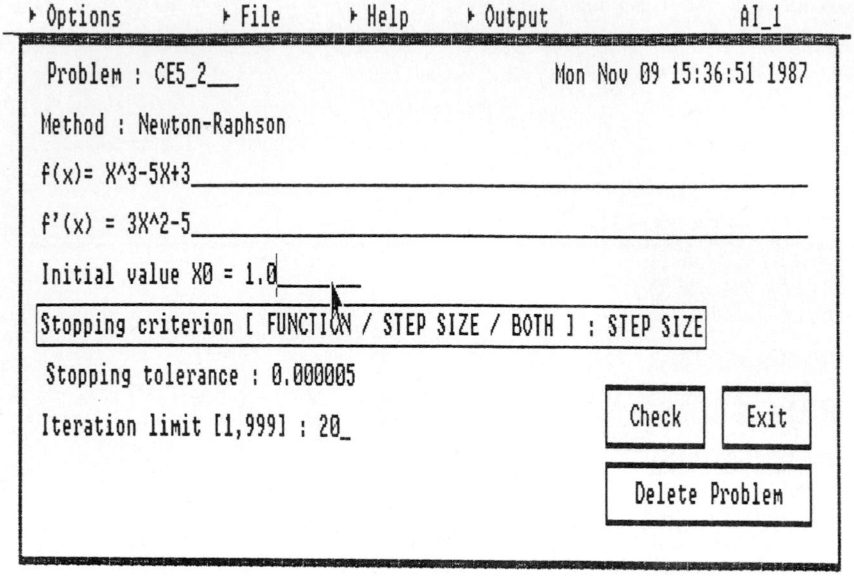

```
 ▸ Options        ▸ File     ▸ Help    ▸ Output              AI_1
┌──────────────────────────────────────────────────────────────┐
│  Problem : CE5_2___                    Mon Nov 09 15:36:51 1987 │
│                                                                │
│  Method : Newton-Raphson                                       │
│                                                                │
│  f(x)= X^3-5X+3_____ │
│                                                                │
│  f'(x) = 3X^2-5_____  │
│                                                                │
│  Initial value X0 = 1.0_____                                  │
│  ┌──────────────────────────────────────────────────────────┐ │
│  │ Stopping criterion [ FUNCTION / STEP SIZE / BOTH ] : STEP SIZE │ │
│  └──────────────────────────────────────────────────────────┘ │
│  Stopping tolerance : 0.000005                                 │
│                                            ┌────────┐ ┌──────┐ │
│  Iteration limit [1,999] : 20_             │ Check  │ │ Exit │ │
│                                            └────────┘ └──────┘ │
│                                            ┌────────────────┐  │
│                                            │ Delete Problem │  │
│                                            └────────────────┘  │
└──────────────────────────────────────────────────────────────┘
```

Figure 4

The problem can be input by clicking the mouse on a particular line to input or edit data. The check option is to enable students to confirm that they have input a complete problem. The 'Solve problem' option will not run a problem unless it is complete.

5. CONCLUSION

We believe that the two ways in which we have used the microcomputer to teach computational mathematics will provide the framework for producing high quality, user-friendly software. Other Open University courses involving numerical methods will also work within this existing framework. We feel that many other institutions could save a great deal of effort if they were to benefit from the developments we have made.

BIBLIOGRAPHY

1. M351: Numerical Computation (Open University Press, Milton Keynes, 1976).
2. M371: Computational Mathematics (Open University Press, Milton Keynes, 1988).
3. M101: Mathematics foundation course (Open University Press, Milton Keynes, 1977).
4. MST204: Mathematical models and methods (Open University Press, Milton Keynes, 1982).

FCM/87 - Educational Computing in Mathematics
T.F. Banchoff et al (editors)
© Elsevier Science Publishers B.V. (North-Holland), 1988

THE LABORATORY OF MATHEMATICS: COMPUTERS AS AN INSTRUMENT FOR TEACHING CALCULUS

I. Capuzzo Dolcetta, M. Emmer, M. Falcone, S. Finzi Vita

Dipartimento di Matematica
Università di Roma "La Sapienza"
Piazzale A. Moro, 2
00185 Rome - Italy

Personal computers has been extensively used in the courses of Calculus and Advanced Calculus at the University of Rome in the last four years. In particular they have been used to illustrate mathematical phenomena as well as to solve problems. This fact obliged teachers to change the contents and the organization of traditional courses. We shall present some recent developments of this experiment showing examples of the results which were obtained and discussing in details some related questions: what kind of organization is the best ? what language is more adapt to the target ? which are the essential features requested for a demonstration software in this field ? how final examinations can be organized ?

1. THE FIRST EXPERIENCE

Starting from 1983 computers have been introduced in the courses of Calculus and Advanced Calculus at the University of Rome. The main reasons to introduce them are related to the importance of computers in modern mathematics: first of all for the possibility they offer to solve difficult mathematical problems whose (theoretical) solution cannot be exhibited in an explicit form, and secondary for the hints that a computer simulation can give in the study of many mathematical phenomena. It is quite important to notice that the technological (r)evolution of the last decade has made possible to use for all these purposes even very small computers, such as personal computers. This new and exciting situation has stimulated many teachers to present various applications of computers in mathematics in their courses starting from the very beginning of university curricula (or even in secondary schools). In fact our experiment has been made in connection with other universities and in particular with the Universities of Paris-Sud Orsay and Leeds as part of a project supported by the Commission for Education of the European Community (see also [1] and [2] in this volume). Starting our experiment we were convinced, and we still are, that every student in any scientific discipline should be able to use a computer for scientific investigations and have an informatic background. This result can be obtained either introducing new compulsory courses on these topics or changing the contents of traditional courses: the quite rigid organization of curricula in the Italian university made easier to follow the second way [3].

As everyone knows, both Calculus and Advanced Calculus are foundamental courses devoted to develop logic-deductive skills while presenting mathematical techniques and results requested in the following courses. Traditionally the organization of these courses is made up by general lectures and practical exercise sessions. The introduction of computer induced some modifications in both of them, since computers have been used to present and visualize mathematical phenomena as well as to solve problems from a numerical point of view.

In the first two years of the experiment we prepared demostration programs to visualize and illustrate some aspects of mathematics using graphic facilities offered by personal computers. These programs constitute the basis for a real *Laboratory of Mathematics* in which you can present and discuss a number of different experiments: functions, sequences, solution trajectories of ordinary differential equations, curves and surfaces can be exhibited to students showing them the wide variety of behaviours and situations hidden in a mathematical definition. The visualization of a series of examples helps students to absorb concepts that can be quite difficult in their abstract mathematical formulation, particularly in the first two years of their curriculum. From this point of view the use of a computer is very practical and effective since, if your program allows an interactive modification of data, you can easily modify the parameters in a given example showing immediately the changes due to the new choice. Students can use this *dynamic blackboard* to make investigations, verifying their knowledge and the validity of their intuitions. At the very beginning it is important to guide these investigations since, expecially in mathematics, answers given by a computer are not always correct. Nevertheless, even wrong answers can be important to solicit and motivate a presentation of some interesting topics related to computer applications in science, namely the representation of numbers in binary and exadecimal form, the use of floating point numbers, the axiomatic of real numbers and the realization of elementary operations on a computer. All these topics were discussed in our courses, in addition to traditional contents.

The main point regarding programs developed for demonstrations in mathematics, is that they must not be so sophisticated to make typical computer errors disappear. Student must be aware of limitations connected with the use of computers in scientific investigation and programs must not cancel mathematical difficulties: an overflow is an overflow and must not vanish by means of a programming trick! It is also very important for the student to have a sufficient background to distinguish his errors from computer errors. This leads to the second applications of computers in our courses.

In the courses of Calculus and Advanced Calculus the accent is on definitions and *qualitative* results which can be proved starting from definitions: convergence, existence, regularity, uniqueness and so on. We decided to treat in our courses also new topics related to *quantitative* results since in many applications quantitative aspects are as important as qualitative ones. Why, for example, focusing the attention on the results of the theory of integration and on methods which lead to an explicit solution of an integral if, in the great majority of practical applications, these methods cannot be used ? Why not give to students some elementary tools to compute a definite integral ? One of the main answers is that the course program is so large that time is not enough to add new topics. In fact a standard course consists of about 120 hours (summing up general lectures and pratical exercise sessions). Nevertheless, our experience has shown that it is possible to treat these topics adding only a small number of hours to the schedule. Students' response to this experiment has been enthusiastic due to their interest in computers applications. The topics which have been added (see Table 1), though quite elementary , are sufficient to face the typical problems introduced during the course.

As it can be seen in Table 1, the total duration of the course was augmented of about 1/7. The first 6 hours were devoted to an introduction to programming and to a quick presentation of the BASIC language. The main target is to drive students rapidly towards an active use of computers postponing the refinements to the discussion of particular problems. In this way students can start programming from the very beginning, trying by themselves to implement numerical methods in order to solve concrete mathematical problems. This active work of programming, which was done in a Laboratory with 10 personal computers Olivetti M-24, is important to make the student understand the power and the limits of numerical methods and is crucial to get him used to the

---

TABLE 1

FIRST YEAR

INTRODUCTION TO THE LOGIC OF PROGRAMMING
(cicles, alternatives, flow-charts) .... 2 h
  BASIC .... 4 h
COMPUTING ERRORS
(rounding, loss of significant digits) .... 1 h
GRAPHS OF FUNCTIONS AND GRAPHIC COMMANDS OF BASIC .. 2 h
METHODS TO LOCATE ZEROS OF FUNCTIONS
(bisection, secant, Newton's methods) .... 2 h
NUMERICAL METHODS FOR COMPUTING DEFINITE INTEGRALS
(rectangles, trapezoidal and Simpson's formulas) .... 2 h

SECOND YEAR

DIRECT METHODS FOR SOLVING LINEAR SYSTEMS
(Gauss, pivoting) .... 3 h
SURFACE PLOTTING .... 4 h
METHODS FOR INTEGRATING ODE
(Euler, modified Euler, Taylor, Runge Kutta) .... 6 h
METHODS FOR NONLINEAR OPTIMIZATION
(gradient, projected gradient) .... 4 h

---

application of computer as a research tool from his first year at university. In the *working sessions on computer* the emphasis is mainly on mathematics rather than on programming. In this direction the choice of a programming language is not very important and BASIC is probably the best choice to minimize the number of prerequisites, considering that many students have already used it on their home-computers. During the sessions at the Laboratory students work in groups of two or three writing their programs and solving exercises: this organization helps the exchange of informations, accelerates the development of their programming skill and contributes to make the entire group more homogeneous, as it has been shown by some tests (see [3] and next section 4).

2. RECENT DEVELOPMENTS

Until 1985 the experiment described in the first section was limited to small groups of students due to the small number of computers available and to the will of testing materials and organization . It must be noticed in fact that the courses of Calculus and Advanced Calculus are attended by 180 and 100 students respectively so that it is quite a big work to organize tutoring and Laboratory sessions for them. During the first three years of the experiment only 30 students for each course took part to the working sessions on computers and followed some additional lectures on the topics illustrated in Table 1.

Starting from 1986 the experiment has been done on a larger scale. The topics of Table 1 were introduced in the official programs of courses and it became possible to use software during the general lectures by means of a computer connected to a tricromic projector. At the beginning of the course students could decide to participate to working sessions on computers. This activity in the

---

TABLE 2

Examples of examination problems:
1. Find all the zeros of the function $f(x) = \cos 3x - \cos 2x + 0.71$ in the interval $[-\pi, \pi]$, with an error less than 0.001 .

2. Find all the real roots of the polynomial $x^3 - 23x^2 + 159.99x - 299.97$ with a precision of 0.001 .

3. Solve the following system of equations: $y^2 - x + 1 = 0$
                                                  $y - \sin x - 1 = 0$,
   computing all the roots with an error less than 0.001 .

4. Find the absolute maximum and minimum points of the function $f(x) = x \sin x$ in the interval $[-1,6]$, with a precision of 0.001 .

5. Solve problem 4. for $f(x) = \int_{-1}^{x} \arctan t \, dt$ in $[-1, 3]$.

6. Solve problem 4. for $f(x) = \int_{0.5}^{x} (\ln t / t^2) \, dt$ in $[0.5, 6]$.

7. Solve problem 4. for $f(x) = \int_{1}^{x} (\sin t / t) \, dt$ in $[1, 10]$.

In all the cases explain and justify theoretically the used method.

Laboratory was not compulsory but those who decided to participate were submitted, at the end of the course, to an additional examination whose result contributed to the final mark. The main reason to leave students free to make this choice is that they enter university with a very unhomogeneous mathematical background: many of them have already enough difficulties following the course and do not want to add an extra work of about 4 hours to their weekly schedule. In fact this is what most of them declared in an anonymous questionnaire (see section 4). Nevertheless 80 students over 180 decided to participate to the Laboratory sessions.

The examination for them consisted in the solution of a particular problem. During this examination, whose duration is 1 hour, the student was alone in front of the computer but he was allowed to use the software developed by his group. The proposed questions were difficult enough to force him to make a precise analysis of the mathematical problems and, eventually, to modify his software in order to overcome the difficulties and reach the solution. They were requested to explain and justify from a theoretical point of view the method they had applied to find the numerical solution. Typical examples of the text of these examinations are presented in Table 2.

In the course of Advanced Calculus (which deals mainly with the theory of functions of several real variables, differential equations and Fourier approximation) the quantitative approach requires very often longer and more complicated programs. Even if it is possible to develop these programs using BASIC language, it can be a good idea to switch to a more advanced and structured language such as Pascal. In particular the Turbo-Pascal version is very appropriate for its simplicity and was therefore adopted in that case. The organization of this course was exactly the same of Calculus and about 30 students decided to take part to Laboratory sessions. It is quite interesting to notice that about 15 of them did not participate in the first year: even if they started programming directly in Pascal without any previous exeperience in Basic, after two months they were at the same level of their colleagues. They worked in groups of two for the whole semester and the final examination consisted of a written report on the numerical solution of a particular problem chosen between those listed in Table 3.

TABLE 3

Examples of problems proposed to second year students:

1. Numerical solution of non-autonomous ordinary differential equations
$$y'(t) = f(t, y(t)) , \quad y(0) = x .$$
2. Graphs of surfaces in assonometric and central perspective.
3. The gradient method for optimization. The projected gradient on a circle and on a square.
4. Fourier series approximation of a given function.
5. Finite difference approximation of the solution of
$$- y'' + q(x) y = g(x) \quad , \quad y(a) = A , \quad y(b) = B .$$

All these subjects were quite advanced for second year students and required a good knowledge of mathematics as well as a good informatic skill. Some of them were not presented in full details during the lectures so that students were also requested to find the appropriate implementation of different methods and link them at the interior of their programs.

A typical example of this kind of work is the Fourier approximation of a periodic function of one real variable. As it is known, if f is a periodic piecewise continuous function on an interval, say $[-\pi,\pi]$, the following sequence

(1) $$f_n(x) = \frac{a_0}{2} + \sum_{k=1}^{n} (a_k \cos kx + b_k \sin kx) \quad,$$

where

(2) $$a_k = \frac{1}{\pi} \int_{-\pi}^{\pi} f(x) \cos kx \, dx \quad, \qquad b_k = \frac{1}{\pi} \int_{-\pi}^{\pi} f(x) \sin kx \, dx \quad,$$

converges to f(x) for any x where f is continuous and converges to

$$\frac{1}{2} [f(x_0 - 0) + f(x_0 + 0)]$$

at any point of discontinuity x_0. If you want to compute and design the Fourier approximation of order n of f you will need first of all its coefficient a_k and b_k, which you can obtain by means of a numerical method for definite integrals, then the approximate values of f at a number of nodes x_i, computed applying (1), and finally you can use these data to plot the approximate function. To have a good approximation in a reasonable amount of time you will need an accurate and fast algorithm to compute definite integrals since it is necessary to compute 2n coefficients. In particular, noticing that the functions sin kx and cos kx oscillate faster and faster when k increases, it would be a good idea to increase, proportionally to k, the number of nodes used for the computation of (2). For the final examination students developed a program which solves this problem efficiently (see Fig.1).

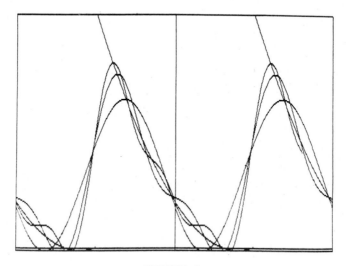

FIGURE 1

Fourier approximations of increasing order of a periodic
discontinuous function of one real variable.

The choice of the programming language is essentially related to the local situation in terms of equipment, to the objectives and to the background that the students already have. Actually PASCAL and BASIC seem to be the more natural choice since they are the most popular general purpose languages and provide graphic facilities, but C and FORTRAN could be preferable if the emphasis is on numerical analysis and the final target is scientific computing. Other languages, such as APL and LISP, are more suitable for courses of algebra and geometry, where the possibility to treat abstract objects instead of numerical data is more valuable. In our experience, anyway, the switch from a language to another can be very rapid if students have a sufficient background.

3. GRAPHIC ILLUSTRATION OF MATHEMATICAL CONCEPTS AND METHODS

As already mentioned, in our experience computers were used in two different ways, both of them important.

Even though the active work of programming in order to solve mathematical problems remains in fact crucial, computers may be also powerfully employed in a classroom to investigate mathematical phenomena, with the help of a tricromic projector or, more simply and economically, of a liquid crystal display connected with an overhead projector.

As well as software is concerned, we have noticed that both ready-made packages (many of them are now available) or simple demonstration programs can fit to the aim of these lectures. The important fact is to have enough interactivity and graphic effectiveness at one's disposal, in order to avoid these lectures to become boring and useless.

What remains fundamental is, anyway, teacher's role, since he has to be in this context the real animator of the *happening* . We think that the computer has to remain a tool in his hands: showing to a passive audience only a series of beautiful images can at the most create astonishment, but it does not help students to understand the theory and the phenomena hidden behind them.

After recalling to the students the basic concepts of the theory previously introduced during the general lectures, the teacher shows with the help of a computer a few simple and known examples (those already seen on the blackboard during the exercises sessions or those usually present in text-books) in order to confirm the theory. Little by little the examples will be more complicated and some counterexamples to the theory can be shown: this helps to focus the essential hypotheses for a property to hold. Moreover it happens very often that a very little change in the data produces a drastic change in the results. In this way every new input encourages the discussion: the teacher can ask the students to propose new examples and to foresee what should appear on the screen and why. These *gymnastics* produce at the end a better understanding of the whole problem.

There are many topics in which this (graphic) approach can be useful. Some of them are standard ones, such as representation of sequences and functions , approximation of functions by polynomial interpolation, Taylor or Fourier series, visualization, after numerical integration, of trajectories of ODE solutions.

In this section we want to describe in particular another aspect: how the graphic aid of computers can help to explain notions and methods of mathematical analysis.

One of the most common objections to this use of computers is that a computer cannot handle concepts such as the limit of a sequence. This is true if we think to determine with a computer the limit value of the sequence, or even to approximate it if we have not yet discussed the qualitative character of the whole sequence (i.e. convergence, monotonicity, boundedness, etc.): as we always explain to first year students, even looking to the first two millions elements of a sequence does not help to determine with certainty its limit. But this machine limitation is the same that our minds have to overcome: every time in our courses we deal with infinity (and it happens very often), we are forced to introduce definitions which are not *natural* in terms of intuition, and which therefore are not easily understood by students. Let us recall the well known definition of limit of a sequence:

(3) $" \lim_{n \to \infty} a_n = L \quad \Leftrightarrow \quad \forall \varepsilon > 0 \quad \exists \ n_o = n_o(\varepsilon)$ such that $\forall n > n_o : \ | \ a_n - L \ | < \varepsilon$

According to our experience, many students still have problems with this definition during their final examination. The use of colour and graphic facilities on a computer allows a better understanding of this concept. It is in fact simple to write a program to represent a real sequence a_n by means of the points (n, a_n) in a system of cartesian co-ordinates on the screen. Then this program, slightly modified, can be used to verify (both graphically and numerically) the convergence of the given sequence to its known limit (or perhaps its divergence). It is enough to use (in a dynamical way !) the definition (3). The main steps will be the following:

- *The choice of* ε. The program asks for an input. It is then clear that ε is the first quantity to be chosen .

- The strip $(L-\varepsilon, L+\varepsilon)$. In the graph of the sequence a coloured strip appears around the presumed limit value L ; ε is easily interpreted as the half-width of the strip and the condition $| \ a_n - L \ | < \varepsilon$ appears to be satisfied by the elements (points) lying at the interior of the strip.

- *Looking for* n_o. The program rapidly computes, if existing, the first index n_o such that all the subsequent elements (up to a given number N) lie within the strip. A greater N can be used to continue this research on a bigger set of indices.

- n_o *depends on* ε. Varying ε, its general influence on n_o can be immediately shown, together with the essential role of the symbol " $\forall$ " (" $\forall \varepsilon > 0$ .... ").

As an illustration of that, Fig. 2 shows, without colours, an output of the program in the case $a_n = (-1)^n / n$.

The same idea can be used to introduce with more efficacy other mathematical notions, such as: infimum or supremum of a sequence, upper or lower limit, continuity and uniform continuity of a function, integral of a bounded function on a compact set, etc.

What these programs perform are, more or less, the same constructions previously sketched by the teacher on the blackboard, but now in a more efficient and dynamical way. The computer then plays the role of a dynamic blackboard, essential complement of the traditional one.

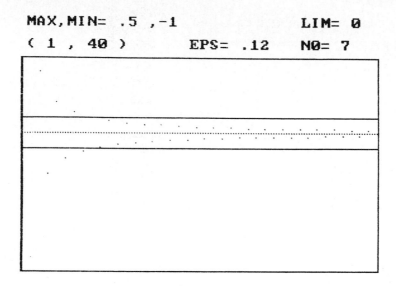

FIGURE 2

Another typical example is the following: a simple BASIC program can allow to draw graphs of functions on four different portions of the screen. This programs reveals itself very useful to show for example the effects of parametric perturbations of a given function. Students can rapidly learn how look like functions as $f(x+p)$ and $f(x)+p$ [p positive or negative], $f(px)$ and $pf(x)$ [p greater or smaller than 1] comparing them with the reference function $f(x)$. This is important in particular at the beginning of first year courses since students do not distinguish yet very well between domain and range of a function and are not familiar with notions such as translations, expansions or contractions. Two examples are shown in Figs. 3 and 4.

Every other case of a one-parameter family of functions $f(x,p)$ can be treated as well, such as exponential or logarithmic functions with different basis, solutions of first order ordinary differential equations, sequences of functions (where $f(x,p)=f_p(x)$, with $p \in$ IN).

Finally we want to mention the way we introduced, through these computer lectures, those topics that in Section 1 we considered as *quantitative aspects of mathematical problems* . Simple numerical methods, such as the methods for computing definite integrals or those for locating and approximating zeros of functions, have been usually explained during the general lectures and the main convergence results have been recalled without proofs. But the best way to make these methods clear seemed to us that of showing in a graphic way their action on different examples.

In the first case (definite integrals), the program uses colours to fill the approximated area computed by the choosen quadrature formula and draws it in superposition to the graph of the given function. Changing the number of points considered by the formula, the effects on the error can be evaluated at least qualitatively (for example approximation from below or from above).

I. Capuzzo Dolcetta et al.

1)intervals 2)parameter 3)points 4)window 5)draw 6)stop

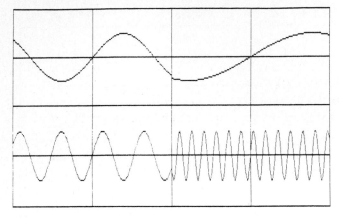

FIGURE 3

f(x) = sin (px) in [-4,4], p = 1, 0.5, 3, 10

1)intervals 2)parameter 3)points 4)window 5)draw 6)stop

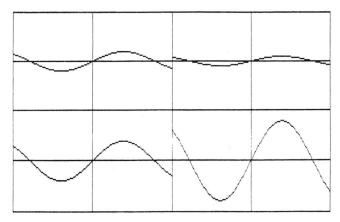

FIGURE 4

f(x) = p sin (x) in [-4,4], p = 1, 0.5, 2, 4

Moreover, since the program aim is only demonstration, when the primitive function is known the error can be also computed exactly (except that for rounding errors, of course).

In the second case (zeros of a function), the program allows to follow the iterative methods step by step on the graph of the given function. Tracing, for example, medium points (for the bisection method), secant or tangent lines (for the secant or the Newton's method) as they are generated by the algorithms, helps to understand how these methods work and when one of them is preferable to the others.

The use of computers that we described appears to be useful in particular for those students who have troubles with abstract reasoning, for whom evidence and intuition remain, at least at the beginning, fundamental. Moreover, these *computer lectures* can provide all the students with a deeper knowledge of many aspects of the theory and, at the same time, with a greater insight in the applications.

4. EVALUATION OF THE EXPERIMENT

An interesting element to evaluate the impact of this experiment has been a questionnaire submitted to all the students at the end of the year. This anonymous questionnaire was divided into two parts. The first one was devoted to general informations: students had to specify their high-school background, the eventual possession of a home-computer, whether they participated to the Laboratory sessions or not, and if not why, and in general their opinion on the whole experience. The second part was a real test made up by nine mathematical questions: students had only 20 minutes to choose the right answer among those proposed. We wanted to ascertain whether (and, possibly, how much) the fact of solving mathematical problems during the computer sessions gave the students an effective advantage.

The results obtained in 1986 were contradictory, probably because the small number of students involved was not statistically significant: the difference between the final average mark of the Computer Group Students (CGS) and that of the Standard Course Students (SCS) was not relevant. Moreover, it seemed that on theoretical questions the CGS were even disadvantaged with respect to the *uncontaminated* SCS (for more details see [3], whose Appendix contains the text of the questionnaire).

For this reason the same test has been repeated one year after on a larger number of students (almost all the students of first year course), and the results have been this time encouraging on the validity of the experiment. We summarize the more interesting ones in Table 4.

```
┌─────────────────────────────────────────────────────────────────┐
│                          TABLE  4                               │
│                                                                 │
│  - Number of students who answered the questionnaire:  152      │
│      Students who own a computer:           36.8 %              │
│      Students who intend to buy a computer:  47.3 %             │
│                                                                 │
│  - Computer Group Students (CGS):   67   (44 %)                 │
│     Standard Course Students (SCS):   85   (56 %)               │
│                                                                 │
│  - Marks :                                                      │
│         General  average  mark    4.13 / 9                      │
│         CGS          "        "      4.71                       │
│         SCS          "        "      3.67                       │
│                                                                 │
│  - Students who passed the test  (more than 4 right answers):   │
│         Global result    40.8 %                                 │
│         CGS              53.7 %                                 │
│         SCS              30.6 %                                 │
└─────────────────────────────────────────────────────────────────┘
```

The fact that almost all the students have a computer or are going to have it soon is not surprising, but it is a datum which cannot be ignored when starting this kind of experiments.

The SCS motivated their non-participation to the Laboratory sessions with lack of time or with the difficulties they already had with the standard topics of the course. Almost all the CGS were satisfied of their participation. Only a few (6%) considered the results obtained limited in comparison with the time needed to reach them.

As it is shown in Table 4, the difference between the results of the CGS and those of the SCS appears now relevant and, what is more inportant, it has been almost the same on every question of the test. As we hoped, the work in groups in the Laboratory seems to have enriched and positively stimulated the students.

This kind of results, together with the students' enthusiasm, encourages us to continue the experiment, since we realize that much still remains to be done.

REFERENCES

[1]   Dechamps, M., A European Cooperation on the Use of Computers in Mathematics, this volume.

[2]   Salinger, D., The Teaching of Mathematics in a Computer Age, this volume.

[3]   Capuzzo Dolcetta, I., Emmer, M., Falcone, M. and Finzi Vita, S., The Impact of New Technologies in Teaching Calculus: a Report of an Experience, to appear on Intern. J. on Math. Education in Science and Technology.

[4]   Capuzzo Dolcetta, I., Falcone, M. and Picardello, M., Utilizzazione dei microcomputers nell'insegnamento della matematica nel primo biennio universitario scientifico, CEE Report of Contract SSV-84-397-I, Rome (1985).

[5]   Cottet-Emard, F., Garcia, F. and Rivier M., L'enseignement des mathematiques par les moyens informatiques en premier cycle universitaire, CEE Report of Contract SSV-84-367-F, Paris (1985).

[6]   Emmer, M., Falcone, M. and Finzi Vita, S., Computers nella didattica della matematica, Videotape, CNR Project "Nuove tecnologie per la didattica", 15 minutes (1986).

[7]   Capuzzo Dolcetta, I. and Falcone, M. , L'analisi matematica al calcolatore (Zanichelli, Milano, 1988).

[8]   Lax, P., Burstein, S. and Lax, A., Analisi matematica con applicazioni e calcolo numerico (Zanichelli, Milano, 1986).

[9]   Marcellini, P. and Sbordone, C., Complementi di Analisi Matematica e programmazione in BASIC (Liguori, Napoli, 1986).

ECM/87 - Educational Computing in Mathematics
T.F. Banchoff et al. (editors)
Elsevier Science Publishers B.V. (North-Holland), 1988

DIDACTICAL ASPECTS OF THE USE OF COMPUTERS FOR TEACHING AND LEARNING MATHEMATICS

Bernard CORNU

Université GRENOBLE 1 – FRANCE
Institut Fourier
Equipe de Recherche en Didactique des Mathématiques et de l'Informatique

The computer changes the knowledge, the teacher, and the pupil, and the interactions between those. Recent developments of didactical research provide tools to improve the teaching and the learning of some mathematical concepts, with the help of the computer. In this paper, we show some changes due to the computer, and we give some examples of a didactical approach of the use of computers.

1. THE COMPUTER, THE KNOWLEDGE, THE TEACHER, THE PUPIL .

Computers are generally used for teaching without a strong control on how they act upon the knowledge of the pupil. Educational research ("Didactics of mathematics") provides tools and results about teaching and learning; it seems that we should integrate these results in the use of computers. Didactics of mathematics studies the pupil's acquisition of mathematical knowledge. It provides a theoretical framework for the study of learning processes. The basic object is the didactical system : the knowledge, the teacher, the pupil, and the interactions and relationships among them. It is usually represented by a triangle.

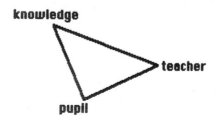

For example, the relationship between the teacher and the pupil includes implicit rules about their behaviour, and these rules, sometimes called "didactical contract", influence the learning. The relationship between the teacher and his own knowledge also influences the teaching, and therefore the learning. But of course the most studied is the relationship between the pupil and the knowledge : how the knowledge is structured within the pupil's mind, what are the mental representations and images of a given concept, which conceptions do the pupils have about a concept, how to act upon these images, conceptions, representations, which are the obstacles to the learning, what is the meaning of the errors, etc... The didactical system has not to be considered as isolated : it is embedded in the society, and in particular in the mathematical society; but a mathematical concept as it is taught is not identical to the concept as mathematicians know it. The mathematical knowledge is generally transformed into a knowledge to be taught, and this transformation is called "didactical transposition". All these questions are usually studied by didacticians, and they try to elaborate situations in classrooms to help the learning.

But how do computers intervene ? Is the computer a supplementary tool among other kinds of tools ? Does it have specific effects ? In fact, it seems that it has not to be considered as an extra component in the didactical system, as an extra partner for the pupil or the teacher. It changes each of the three components of the didactical system, and what we shall examine is this new triangle, where the knowledge, the teacher, and the pupil are changed by informatics and computers :

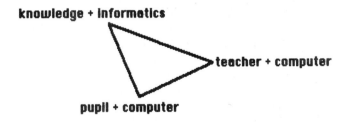

knowledge + informatics

teacher + computer

pupil + computer

We shall describe what are these new components : the "new knowledge", the "new teacher", and the "new pupil". We shall see how informatics and computers change the mathematical knowledge, and the knowledge to be taught, how they change the teaching, and how they change the learning.

2. KNOWLEDGE + INFORMATICS .

Not only computers, as materials and tools, influence mathematics, but mostly informatics as a science with its own concepts, its own methods. The fundamental concepts of algorithmics have a strong influence on mathematics and on mathematical activity. Some mathematical **concepts** evolve : the concept of a number, the concept of a variable, the concept of a function, etc. Other concepts appear or are developing, in relation with computer science concepts (for example, induction and recursion, concepts associated with data structures, etc.).

Mathematical activity evolves :

- Simulation, modelling, visualization, experimentation take more and more place within mathematical activity. The computer is a tool for experimentation, and a tool for elaborating conjectures.

- The computer plays a role in problem solving. It helps for long computations, for computation with great numbers, for symbolic computation. Graphical possibilities of the computer can help to solve problems in several domains, expert systems and artificial intelligence can provide a help for discovering proofs and for automatic proving. The computer can be considered as a partner for problem solving.

- The concept of **proof** is changing, as also the activity of proving. New kinds of proofs appear, and the question is posed whether these are proofs or not, in the sense that mathematicians accept them as proofs or not. We now see proofs using large computations, proofs using large numbers, proofs using a large number of cases, "algorithmical proofs", i.e. proofs based on the effectiveness of an algorithm, and also automatically elaborated proofs, using expert systems. The computer can be also considered as a partner for proving.

Algorithms see their status change within mathematics. They become more important, because one now prefers constructive proofs rather than existence proofs, so that one has to give algorithms to exhibit mathematical objects. Algorithms are becoming a subject of study by themselves : basic concepts of algorithmics are studied, new algorithms are elaborated, proving and evaluating algorithms are important questions. There appears an "algorithmic approach" of mathematics. The computer also modified the balance between pure and applied mathematics, and changed the development of some branches, bringing new developments and new problems. Dynamic systems and fractals, large prime numbers, symbolic computation, are typical examples. But there are many others, within each branch of mathematics.

These changes within mathematics lead to changes in the content of teaching: mathematics to be taught is changing. One often asks the question : "What can I do with a computer in my classroom ?" In fact, this is not the main question. One should ask : "With the development of computers and computer science, what mathematics should be taught, and how that mathematics should be presented ?" [6] . This raises for instance the question of the place of discrete mathematics in the curricula, the question of the place of logic, the question of an experimental approach of mathematics, the question of a constructive presentation of mathematics, the question of algorithmical mathematics and of the role of algorithmics in mathematics teaching. The mathematical needs of society are changing; one has to take this into account; in many countries, new curricula are elaborated and recommendations for the future are stated.

3. TEACHER + COMPUTER .

In this part, we consider the computer as a tool for the teacher, as a mean for teaching. There are many different ways how to use the computer :

The computer as an "improved blackboard"; one computer in the classroom, used by the teacher, to demonstrate phenomena and mathematical objects, to illustrate, to simulate, to visualize, using the numeric and graphic possibilities of the computer.
Individual computer, in the classroom. Each student or each group of students has a computer and can use it for programming or using packages. The computers can be networked, allowing some communication between them, and specially between the teacher's computer and the students' ones.
Computers in specially equiped rooms, for "practical works", to introduce some concepts before they are taught, or to illustrate them and put them in action after teaching.
The computer as a resource, in a special room, where the pupils can go for exercising, training, testing, or having complementary tutorials.
The computer at home : More and more students have a computer available at home. This has to be taken into account by teachers !

According to these different ways of use, the computer plays many different roles. Fraser and Burkhardt [5] mention the computer as a manager, a task setter, an explainer, a counsellor, a fellow pupil, a resource. But to each of these roles must correspond suited software. There exists many kinds of teaching software: languages, and languages specially adapted for mathematics; software to illustrate, demonstrate, experiment; tutorials; symbolic systems; spreadsheets; "microworlds";...

Hardware and software are not enough. One has to build activities, within a pedagogical strategy, so that the use of such hardware and software is really a help for the student. Hardware and software do not automatically provide knowledge; we need a pedagogical environment. There remains an important work to be done : a work of "didactical engineering".

4. PUPIL + COMPUTER .

One speaks a lot about the computer as an aid for teaching This is the viewpoint of the teacher. Let us take now the viewpoint of the learner: Does the computer have a role as an aid for learning ?

In a sense, the computer extends the brain of the pupil, and modifies the pupil's skills. More and more the computer acquires the skills the pupils were asked to have. It was the case with pocket calculators and numerical computation (how many people claimed " they will no longer be able to compute by themselves !"); it begins to be the case with symbolic computation ("mumath", or the new calculator HP 28 C), and with problem solving, reasoning, and proving. Does the computer replace the pupils' skills ? We have to give the pupils the skill to use computers in a good way. More than the pupils' skills, the computer changes the range of problems we are able to solve and ask the pupils to solve.

In order to make the computer be a true help for the learning, one has to insert it in the mechanism of learning We must analyse which knowledge is produced with the help of the computer. This needs a didactical approach, based on the results of educational research, and using the tools that didactics provide.

Let us see a few examples.

4.1. Many people studied how the knowledge is structured within the mind of a pupil, and gave models or representations of the cognitive structure of a pupil. On a given concept, pupils have conceptions, even before any teaching. Such conceptions are then influenced by the teaching, but they are not purely replaced by "true conceptions". All these conceptions, mental representations, intuitions, examples, images, definitions, theorems, etc., are organized in what David TALL [7] calls the "concept image". The teacher can act upon this concept image, to complete it, to enrich it, to destroy contradictions, and to enlarge it to all the conceptions that are necessary for using the concept. David TALL suggests to elaborate objects able to act upon the concept image : These objects are "generic organisers". They are a kind of microworlds, within which the pupil can manipulate different aspects of the concept, and objects related to the concept. David TALL designed such generic organisers. They are graphical packages, for teaching the main concepts of calculus, and they are designed in such a way that the pupils' concept image about functions, gradients, limits, etc., can be improved.

4.2. An other approach is the epistemological and genetic study of a concept. Ed DUBINSKY [3] analyses the different steps in the acquisition of a concept, and describes which kind of abstraction is necessary at each step. He gives the organisation of these steps, and obtains the "genetic decomposition" of mathematical concepts, which is a tree of knowledge and skills to acquire. Such a decomposition is then used to elaborate didactical situations, able to help the student in the paths of the decomposition. The different programming languages can influence and help mathematics learning. Ed DUBINSKY uses the language ISETL, which is very close to mathematics. A part of the learning of mathematics is transferred to the learning of the language : programming in ISETL makes the student manipulate mathematical objects, and pass some steps in the genetic decomposition of a concept. Tests have been passed by students, showing that activities with this language improve the learning of some mathematical concepts.

4.3. Important obstacles in the learning of a concept are epistemological obstacles. Such obstacles constitute a part of the concept; it is necessary to encounter them in order to acquire the whole concept. They consist of a knowledge, which has been efficient to solve a certain range of problems, but which becomes insufficient to treat a new kind of problem. Their previous efficiency makes them strongly stable for the pupils. To overcome them, it is necessary to destroy the previous knowledge, and to build a new knowledge, suited to the new problem. Such obstacles have been studied by G. BACHELARD. Most of these obstacles can be discovered in the history of the concept, because they have yet been encountered by mathematicians along the elaboration of the concept. Some activities can be designed in order to overcome epistemological obstacles, with the help of the computer. It needs first a strong epistemological study of the concept, and then a didactical study, elaborating situations and problems in which the new knowledge is the only to be efficient.

4.4. The errors of the pupils are generally logic consequences of their knowledge, and if one studies the different possible ways the knowledge about a concept can be structured, which conceptions can occur, then the errors of a pupil allow the teacher to discover the knowledge of this pupil. Errors are symptoms of the cognitive structure of the pupil, and the analysis of the errors is a tool to act upon the pupil's knowledge. But this is quite difficult and needs time. The computer may help the teacher for that, and experiments exist where the errors are analysed and different exercises are proposed to the pupil in order to reajust his knowledge, according to his own cognitive structure.

4.5. The computer can also be used as a partner for problem solving. Some interesting experiments have been elaborated, using expert systems or logic programming. For example, in Rennes [8], they designed a package for problem solving and proving in elementary geometry, with different modules : help to understand the problem; help to analyse the figure; help to find a solving strategy; help to solve the problem; help to prove the result. Starting from the didactical study of a concept and its teaching, the elaboration of pedagogical products which are efficient and reliable is a difficult work of "applied didactics". The computer is a tool for this, but not more than a tool. It has no automatic effect on the knowledge, and must be used in a suited way, within a strategy. There are many different ways to use the computer, and this diversity must be taken into account. More and more, a global pedagogical strategy is needed, along a period of time, integrating the computer with other pedagogical tools.

Didactical engineering consists of techniques and methods to elaborate appropriate teaching strategies. It needs first a didactical study (this is the role of researchers), then designing didactical situations, and elaborating appropriate software. But elaborating good software and elaborating strategies how to use efficiently this software are two different things.

5. PARTICULARITIES OF THE POST-SECONDARY LEVEL .

The uses of computers seem to be quite different at the primary, secondary, and post-secondary levels.

At the primary level, the most used software is LOGO. Programming activities and geometric activities are designed for pupils; many experiments have been done, and the cognitive aspects of such activities have been studied a lot.

At the secondary level, there exists a large number of educational software, designed by teachers, without strong didactical background. Such a software may be efficient when used by its author, with his pupils, but is very difficult to "export". Such a software is often based on the relationship the teacher has with his own knowledge, more than the relationship between the pupil and the knowledge. Such a software is often difficult to integrate in a global strategy, taking into account the time. This leads to a rather limited use of computers in secondary schools, although many computers are available there.

At the university, within mathematics teaching, the computer is used in many different ways :

- Teaching computer science, programming, algorithmics. The most used languages are BASIC and PASCAL, and sometimes LISP or PROLOG.
- The same usual programs are written year after year by many teachers and students. For a better conservation, and for a modular way of programming, there appear libraries of programs and procedures (for example, see [9]).
- Some data-banks for exercises and problems are elaborated.
- There are many packages for simulation, visualisation, specially graphic packages.
- Some "tutorials" are used, for example for calculus and for linear algebra.
- Sophisticated systems such as symbolic systems, spreadsheets, or specialized languages for mathematics, are more and more used. The

didactical experiments designed at the university level often use such software.

- Very often, students are asked to perform a personal project about a mathematical concept or problem, writing a program and running it. These project are generally of high quality, but few of them are re-used.

Generally speaking, one can say that there is very few circulation of ideas and experiments, few conservation of experiments, and few scientific control on what is produced : software and pedagogical environment.

6. TEACHERS' TRAINING.

Changes in mathematics and its teaching under the influence of computers are unavoidable. So it will soon be necessary to have all the teachers trained to the use of computers, and not only enthusiastic ones, as it is the case at the moment. Teachers' training is a key-point, and several aspects have to be taken into account in this training :

- Computer science influence not only the teaching, but also what is taught. Mathematics changes, mathematics to be taught changes.

- The scientific and technological evolution is so that it is not possible to train teachers definitely, for all their carrier long. It is necessary to give them an ability to evolve and adapt.

- It is not enough for teachers to have a knowledge; they need a specific training how to transmit the knowledge. With the computer, the role of the teacher... and the role of the pupil, are changing.

- In-service training is particularly important, because most of the teachers who will teach within the next twenty years are already in service !

- There is too much "amateurism" in the use of computers for mathematics teaching, and specially in the software. It is necessary to train teachers to the diversity of possible uses of the computer, and to the impact of this tool on the learning.

- Teachers need information and elements for evaluating and using the numerous hardware and software which are available.

All these points must be included in the reflexion how to train teachers, and training is a necessary condition for the changes in mathematics teaching to be efficient. The computer changes mathematics, mathematics teaching, mathematics learning; but first, minds must change, and computers do not automatically change the minds !

REFERENCES :

[1] B. CORNU, Logiciels et réalisations informatiques pour l'enseignement

des mathématiques pour l'enseignement des mathématiques dans l'enseignement supérieur, Gazette des Mathématiciens, n°30, Avril 1986, pp. 147-164.

[2] B. CORNU, C. ROBERT, Mathématiques et Calculatrice programmable (Magnard, Paris 1983).

[3] E. DUBINSKY, Using Computer experiences to implement a piagetian theory of learning mathematical concepts, à paraître.

[4] A. ENGEL, Mathématique et Informatique (CEDIC/NATHAN 1985).

[5] A.G. HOWSON et al., The influence of computers and informatics on mathematics and its teaching (Cambridge University Press, ICMI Study Series, 1986).

[6] A. RALSTON, Informatics and the teaching of mathematics in universities, to appear in the proceedings of the IFIP W.G. 3.1 working conference (Sofia, May 1987), North-Holland.

[7] D. TALL, Building and testing a cognitive approach to the calculus using computer graphics (Ph.D. thesis, Warwick Univ., Coventry 1986).

[8] Equipe de Recherche de RENNES (CATEN et IREM), Informatique et ingéniérie didactique (Publ. IREM de Rennes, 1985).

[9] Système MODULOG - Logiciel de l'ALESUP (Publ. IREM Aix-Marseille, 1987).

A EUROPEAN COOPERATION ON THE USE OF COMPUTERS IN MATHEMATICS

Myriam DECHAMPS
Département de Mathématique
Université de Paris-Sud
91405 Orsay Cedex (France)

ABSTRACT. We will describe how the use of computers has been organized in the teaching of mathematics in first and second year courses at the Orsay scientific center of the Paris-South University. We also show how this teaching benefits from cooperation with the departments of mathematics of the universities of Leeds (England) and Rome (Italy), financed by the European Community Commission, through the Education Cooperation Office.

1. INTRODUCTION

Many studies have been devoted to the influence of computers and computer science on Mathematics and their teaching. In particular, the I.C.M.I. (International Commission on Mathematical Instruction) has published substantial documents on this subject [1], and the discussion has been carried on in several conferences ([2], [3], [4], [5]).

In this paper we have chosen another aspect of this question : we will describe in a precise context (the Orsay scientific center of the Paris-South University) how the use of computers has been organized in the teaching of mathematics in the first two years courses.

There is often a disparity between the results of pilot experiments, which are carried out by devoted and qualified teachers on limited numbers of voluntary students, and the institutional setting up of a new work method for all students and teachers. ([6] gives a good analysis of the difficulties and "*perverse*" effects of this transition from the experimental stage to the institutional stage). We think that the analysis of these two models of situations can be of great help.

What are the characteristics of the Orsay center ? Let us say first that in France the university curricula that lead to the "*maîtrise*" take four years : the first two years constitute the "*premier cycle*", the following two years the "*second cycle*". The "*troisième cycle*" is a research-training program, which lasts, at present, about 3 to 6 years.

We will concentrate on the first two years at university : it is at this stage between secondary studies and specialized scientific studies that the choices concerning the use of computers are most urgent : the initial training of students in computer science is only done to a small extent by secondary education ; on the other hand, the widespread use of micro-computers in everyday life demands that universities produce technicians, teachers or scientists who can master the use of computer science in their specialities.

In France the teaching in the first cycle is multidisciplinary. For scientific studies, it includes, in the first year, at least three subjects chosen from among Mathematics, Physics, Chemistry and Biology. Before 1985, the Orsay Computer Science Research Laboratory staff had teaching duties only in the second and third cycle. So, it was the task of the teaching staff of the fundamental subjects to organize the use of computers in the first cycle.

The first experiments are fifteen years old, they were developed as optional teaching, supervised by physicists or chemists. Mathematicians at first only contributed to these options ; then, about five years ago, they introduced their own options. The reasons for this late start are, in my opinion, multiple and related to the context of Orsay : physicists and chemists were confronted earlier than pure mathematicians with the use of computers for their research ; applied mathematicians are not numerous enough and concentrate their efforts above all on the second and third cycle ; the ratio : number of available teachers/number of necessary teachers, in Orsay university, is higher in physics and in chemistry than in mathematics (it was about 171 %, 140 % and 92 % respectively in 1984, according to ministerial norms).

The reform of the first cycle started by the center of Orsay in 1984 has helped the teaching staff of each fundamental subject to specify their aims and their responsabilities concerning the initial training of students in computing. In this paper, we describe the organization chosen from this date on, as far as mathematics is concerned, after a quick description of the characteristics of the center of Orsay, of the department of mathematics and of the teaching in the first cycle.

For mathematical studies one of the great assets has been the cooperation started in 1983 with the departments of mathematics of the universities of Leeds (England) and Rome (Italy), financed by the E.E.C. (Commission des Communautés Européennes), through the Office of Cooperation in Education. This cooperation made it possible to take advantage of the experiments carried out in each university, and the available didactic material. These last two years, aid granted by the E.E.C. for setting up a *"common program of studies"* between the three universities involved, has allowed exchanges of students and teachers for periods of two to six weeks. These exchanges have been a privileged tool to widen each partner's knowledge of the use of computers.

The organization of mathematics teaching in the first cycle in the universities of Leeds and Rome is rather different from that in Orsay. In particular, the teaching is managed more directly by the departments of mathematics ; in Leeds, applied mathematics takes up a bigger part of the curriculum. However, as far as the use of computers is concerned, the aims and the pedagogical choices are rather similar. A European cooperation in this sector has proved quite positive : it allows the necessary adaptations in a rapidly changing field. It also serves as a framework for reflection on the impact of the generalization of computers on the contents and the form of mathematics teaching. This is perhaps one of the most important issues for the coming years : the choices made in this domain will condition the usefulness, the interest and the vitality of university mathematics teaching.

We will conclude this descriptive account with some comments made by the teaching staff involved in the use of computers in Orsay on the problems of teaching with which they are confronted.

2. THE ORSAY SCIENTIFIC CENTER

Situated 25 km south of Paris, it is one of the great scientific centers in the Paris region, with its 15.000 people : 11.000 students, 830 teachers (also involved in research activity), 800 research workers (CNRS), 2.500 engineers, administrative staff and technicians.

The scientific university first cycle is intended for students with a scientific *"baccalauréat"* (the final exam for secondary studies). There is no other entrance requirement, the number of students is limited by the capacity of individual universities (about 1.500 students in Orsay). The first cycle studies take two years and lead to a diploma in general university studies (DEUG) or to a diploma in scientific and technological university studies (DEUST).

The students who have passed the DEUG, and some of the students who have passed the DEUST, can carry on their studies within the second cycle, which lasts two years and get a "*maîtrise*". They can also sit for competitive exams organized by institutes for engineering studies or by administrative organizations, or they can take a year's training which leads to a university diploma in technology.

At the end of the second cycle, the students who have chosen secondary teaching as a profession can prepare for different competitive examinations.

Those who have chosen university teaching or research can prepare a "*diplôme d'études approfondies*" in one year (DEA), then a "*nouvelle thèse*", which takes two or three years, and later "*l'habilitation à diriger des recherches*".

3. THE ORSAY MATHEMATICS DEPARTMENT

At the beginning of the academic year 1986-1987, the department had 125 teaching posts (involving research). Usually the professors (25 % of the teaching staff) deliver theoretical lectures ("*Cours*"), and the other teachers are in charge of exercise sessions ("*Travaux Dirigés*").

The department of Orsay contains five research groups which are associated with the CNRS (National Council for Scientific Research) : Harmonic analysis, Differential topology, Arithmetic and algebraic geometry, Numerical and functional analysis, Applied statistics. The teachers of Orsay belonging to the applied mathematics groups represent less than 20 % of all the teachers.

In 1986-1987, the teaching of mathematics reached : in the first cycle, about 3.000 students ; in the second cycle, 500 students ; in the third cycle, about 100 students.

The department of mathematics uses university computers for first cycle teaching (see § 5). Since 1986, it has its own computers for second cycle teaching : 30 Persona 1600, equipped with fixed disks, and 9 printers.

The applied mathematics groups had their own computers for research and teaching in the third cycle for a long time. Their present equipment consists of :

Statistics group : a VAX 750 with 4 terminals, 7 IBM PC compatible micro-computers (two of these are equipped with Matrox cards which help to drive graphic screens), and 4 Victor 9000 micro-computers.

Numerical analysis group : a Bull SPS 7 mini-computer equipped with a 68 020 and 8 terminals (one of these has a graphic screen) and 6 Persona 1600 micro-computers.

In pure mathematics, the computer equipment is more recent : 6 MacIntosh micro-computers, 5 IBM PC compatible micro-computers and a Bull SPS 7 mini-computer reserved for word-processing.

In addition to these, 11 micro-computers are used for secretarial and library management ; the laboratory is connected to the big computers on campus (UNIVAC 1100 and IBM compatible NAS) and soon an Ethernet network will be built inside the laboratory.

4. ORGANIZATION OF FIRST CYCLE TEACHING

In 1984-1985, the Orsay Center started a reform of the first cycle. The annual system has been replaced by a module system, which allows progressive orientation of students. New teaching units (preprofessional and remedial) have been created, to combat a high failure rate : only one student in three used to obtain the DEUG in two years.

The first cycle is organized in four modules, of 14 weeks each. This is not a semester system : at nearly all levels, repeating is annual and in most cases from February to February of the following year.

4.1. ORIENTATION MODULES M0 : they recruit, in the first semester of the first year, about 1.500 students. Computers are not introduced at this level : attemps in this direction have proved negative, because some students were then tempted to choose their module according to the available computer options and not their subsequent choice of a major subject.

At the end of the module, an orientation council weighs up the results obtained by the student and allocates him to one of the following courses :

4.2. M1-M2-M3 MODULES : These fundamental and multidisciplinary modules associating two or three principal subjects (or more) lead to the DEUG diploma.

The M1-M2 modules (from February of the first year to February of the second year) are organized into various streams (*"filières"*), each with a major subject. In each stream, the major subject teachers are in charge of computer use, which is introduced mainly at this stage, in the form of options with various contents, aims and equipment. Under certain circumstances, students can change stream at the end of modules M1 or M2.

The M3 modules (second semester of the second year) are diversified, according to the different post-DEUG streams ; in addition to the fundamental teaching, they include either options, probationary periods inside or outside the university, or projects, with oral and written reports, on a subject chosen at the beginning of the semester and prepared with the possible help of a tutor.

4.3. P1-P2-P3 MODULES : these preprofessional modules associate a technological speciality with more general teaching and lead to a DEUST diploma, which allows insertion into professional life or possibly further studies. They include probationary periods in factories or work under contract.

There are at the present time three DEUST of this kind in Orsay (Electronics and micro-computer science, Energetics and energy conservation, Biotechnology) and a P3-Laser module which can follow a M1-M2 block.

Together these modules can take about 100 students each year. Mathematicians only do a few hours of teaching here and, because of a teaching staff shortage, they have been unable to propose a specific DEUST (for instance, to train statisticians for the tertiary sector). Computer science is an important subject in the DEUST but it is not taught by mathematicians.

4.4. REMEDIAL MODULES N1-N2 (*"mise à niveau"*) : the students who, at the end of the orientation modules, have not acquired the minimum knowledge required for a M1 or P1 module are guided into a remedial stream (from February to February of the next year). In this stream a DEUG or a DEUST can be obtained in three years.

Mathematics plays a major part here, in proportion to the high rate of failure in the subject (10 hours a week for the sections with mathematics, physics and chemistry, 6 hours for the section with biology and chemistry as major subjects).

Computer use appears here only as an option, supervised by physicists, and recommended for students who aim at computer science studies, as technicians or at a higher level.

For the first three years, the results of the orientation process at the end of the orientation module have been approximately the following :

M1 (DEUG) : 65 %   P1 (DEUST) : 6 %   N1 ("*mise à niveau*") : 29 %

5. THE FIRST CYCLE COMPUTER LABORATORY

This laboratory was created in 1982-1983, under multidisciplinary management, to provide first cycle students with initial training in computer science. It is also used for training courses for members of the administrative or teaching staff on the campus, for further education for adults, and it contributes to the computerized management of the first cycle.

The laboratory is open from 8 a.m. to 7 p.m. : it is reserved for teaching between 9 and 12 and between 2 and 5 p.m., and available for self-service the rest of the time, with reservation on a schedule board.

The use of this laboratory for teaching mainly covers three aspects : programming training, C.A.L., and software use.

Programming training takes place in options organized by teachers of the major subject in each stream and generally concentrated in the M1 and M2 modules described in the preceding paragraph. Five years ago, BASIC was sthe leading language (80 % of the options), then came FORTRAN and PASCAL. The evolution is striking : in 1987-1988, 80 % of the options will use PASCAL in the TURBO version. The swiftness of this new version has made it possible for beginners to take advantage of the formative aspects of this language.

Computer-assisted learning is at the present time used in biology, on software packages created by a group of teachers for the teaching of genetics. The mathematical software created four years ago (formal calculus of primitives) had been written in BASIC and is not used for the moment in the mathematical options. A group of chemistry teachers have carried out an original experiment : C.A.L. software packages created by students within the framework of an option are used by students of the following years.

Commercial packages are used mainly in physics in the second year. This is a trend which will develop if the number of pre-professional modules in the first cycle is increased : firms require employees who are familiar with professional software (word processing, spread-sheets and file management).

The laboratory owns the following equipment :
- 24 Apple II micro-computers ; 8 belong to the first cycle and the others are long term loans ;
- 28 Micral micro-computers (9 are on loan), that function under CPM
- 24 Hector micro-computers (11 are equipped with disk drives)
- 24 Logabax Persona 1 600 micro-computers, that function under MSDOS
- 1 Olivetti M 24 micro-computer
- Access to the consoles of the Univac terminal (11 hours and a half in 85-86)
- about 20 printers (with connection boxes allowing each machine to be
- connected to the printer in turn).

At the present time, this equipment is adequate for the needs of teaching but soon will come the problem of renewing the machines (in particular the Hector microcomputers, because BASIC is used less and less) and it is becoming necessary to invest in peripheral material : plotting tables and digitalizers.

Concerning the teaching staff, a problem to be solved is that of the

assistance required by students who want to learn computer science and who
attend a module where no computer teaching is planned.

Concerning the premises, the numbers of micro-computers per room (from 5 to
8 machines for 10 to 16 students in each room) offers acoustic advantages but
increases the difficulties for introductory lessons : generally 2 or 3
teachers are needed to help the 30 students in one group but the departments
assign only one, because of staff shortages.

6. FIRST CYCLE TEACHING IN THE MATHEMATICAL STREAM

The mathematical stream begins with the second semester of the first year.
It is intended for students who wish to carry on studies in mathematics or
in computer science, possibly in physics, or to prepare competitive
examinations for engineering institutes.

This stream contains three modules of 14 weeks of teaching each, with the
following time-tables (T.D. stands for "*Travaux Dirigés*" i.e. exercice
sessions, and T.P. for "*Travaux Pratiques*" i.e. practical work in laboratory):

6.1. M1A MODULE : second semester of the first year
Mathematics : 4 hours lectures, 5 hours T.D.
Physics   : 3 hours lectures, 3 hours T.D. and 2 hours T.P.
Chemistry   : 1 h. and a half lectures, 1 hour and a half T.D.
Option   : 3 hours. The students can choose between 3 options :

- computer science and mathematics on micro-computers (student
  capacity : 90 to 150)
- computer science applied to mathematics, on a main frame (60
  students)
- preparation to competitive examinations for institutes of engineering
  studies (30 students).

6.2. M2A MODULE : first semester of the second year
Mathematics : 4 hours lectures, 6 hours T.D.
Physics   : 3 hours lectures, 3 hours T.D., 3 hours T.P.
Option   : 3 hours. Mathematics and computer science.

6.3. M3 MODULE, M1 and MP SECTIONS : second semester of the second year.

MI SECTION :
Mathematics   : 4 hours lectures, 6 hours T.D.
Computer science : 4 hours lectures, 4 hours T.D. and 4 hours T.P.
Option   : 3 hours. In 1986-1987 : preparation for the second
                   cycle of mathematics.

MP SECTION :
Mathematics : 4 hours lectures, 6 hours T.D.
Physics   : 3 hours lectures, 3 hours T.D. and 4 sessions of T.P. in the
              semester

The mathematics syllabus of these modules is approximately as follows :
M0 : Continuity and differentiability of real functions. Sequences.
     Polynomial expansions. Integration. Differential equations.
M1 : Linear algebra. Series with positive terms. Geometry of curves.

M2 : Quadratic forms and Euclidian structures. Differential systems. Sequences and series of functions.

M3 : Fourier series. Holomorphic functions. Functions $f : R^n \longrightarrow R^p$. Implicitly defined functions. Surface integrals.

For the previous two years, modules M1 and M2 formed a unit as far as examinations and repeating were concerned. Experience has shown that the analysis syllabus of M2 and M3 is the most difficult to assimilate. Now students may repeat either M1-M2 (from February to February) or M2-M3 (the second year).

The M3 module only contained the MI section at the start. The MP section, which follows the scheme that existed before the reform, has been introduced because some students seemed to prefer this mean of access to the second cycle of mathematics.

The number of students registered in M3 MI in February 1986 represents 8 % of the total number of students in the orientation modules two years before. Among the 500 students in the orientation module with mathematics as their major subject in 1984, 200 obtained a DEUG or DEUST diploma in 1986. Among those who graduated, 13 % have registered for the second cycle of mathematics, 15 % for the second cycle of computer science, 50 % for other cursus and we don't know about the other 22 %.

A substantial mixing of students occurs at the end of the first cycle : in 1986-1987,the first year of the second cycle in Orsay received 239 students in Mathematics and 141 in Computer science.

7. THE MICRO COMPUTER OPTIONS IN THE MATHEMATICAL STREAM

The options aim at illustrating and clarifying the mathematics syllabus to obtain a better understanding of the concepts. They provide introduction to the programming methods that are now indispensable tools for all scientists and into basic computing techniques on computers.

The characteristics of the options offered in the first and second year are similar :

- the same teacher is in charge of the T.D. both for mathematics related to the theoretical lectures, and for computer activities (3 hours a week).

- the students work in pairs on a computer (Micral 90-50 or Persona 1600). They must hand in for marking written reports of their activities (4 or 5) on compulsory themes that are chosen by the teacher.

The mark for the option is the average of the marks obtained for the reports.

- the students can start one or several optional themes according to the remaining time and their inclinations.

- the programming language is PASCAL, in the TURBO version. No preliminary knowledge of a programming language is required. A course in PASCAL is given at the beginning of the option (3 two hours lectures), followed by some introductory practicals on computers.

- the themes tackled with the computers are either examples complementary to the mathematics course, or an approach to new subjects. The students receive stencilled notes, containing, for each theme, a summary of the necessary mathematical notions, hints about and explanations of the algorithms used and the underlying computing problems, and exercices ([7], [8]).

8. WORK THEMES IN THE FIRST YEAR (M1 module)

8.1   The limits of the computer : problems related to the loss of significant digits, order of summation, continuity and differentiability of functions.

8.2   Graphic representation of functions, exercices on centering and scaling.

8.3   Sequences and iterations, fixed point theorem.

8.4   Approximate numerical solution of equations using classical methods : interval-halving, sweeping, linear interpolation, Newton method ; Horner schema for polynomials.

8.5   Taylor's formula, as an approximation formula.

8.6   Approximate numerical integration : rectangle, trapezium, Simpson, Newton-Cotes rules. Comparison of convergence rates, evaluation of errors.

8.7   Solution of linear equations systems with Gauss method, matrix inversion.

8.8   First order differential equations : Euler's method, tangent field, solution curves.

9. WORK THEMES IN THE SECOND YEAR (M2 MODULE)

9.1   Finding the eigenvalues and eigenvectors of a real symmetric matrix using Jacobi's method. Application to statistics : analysis into principal components. Physical interpretation : inertia matrix, principal inertia axes (5 to 6 computer sessions).

9.2   Orthonormalization of a vector system using Gram-Schmidt's algorithm; finding the rank of a vector system, orthogonal projection of a vector into a subspace. Application to statistics : multiple linear regression and least squares method (3 to 4 sessions).

9.3   Tracing of the solution curves of a differential system :

$$x' = p(x,y) \qquad y' = q(x,y)$$

using the Runge-Kutta method. The starting point of the curve is fixed either by moving the graphical cursor on the screen or by giving its coordinates (2 to 3 sessions).

9.4   Fourier series of a periodic function : calculation of Fourier coefficients using numerical approximate integration, simultaneous screen display of a function and of a partial sum of its Fourier series expansion, study of the convergence rate and the behavior in the neighbourhood of points of discontinuity (3 sessions).

9.5   (optional) Brownian movement simulation in the plane, using centered random variables and the central limit theorem (without demonstration). Representation of the Orstein-Uhlenbeck process (1 to 2 sessions).

The themes presented in the second year are less numerous, because they require more work than the themes in the first year, for programming and contents : the theme on Fourier series is a first glance at the theory which will be dealt with during the following semester, the applications to statistics are a first approach to this field.

In the first year, there is another computer science option applied to mathematics, on a mainframe. This option uses the consoles connected to one of the computing centers of the university (UNIVAC 1100) and develops the algorithmic aspect : graph theory, cryptography, dynamic programming, optimization.

We have concentrated our description on the micro-coputer options because they are closely related to the mathematics syllabus and they have benefited from the European cooperation which we will outline in the next section.

10. EUROPEAN COOPERATION ON THE USE OF COMPUTERS IN MATHEMATICS

Although teaching in the first cycle is organized under different forms in the different countries, the questions inherent to the undertaking of the initial formation of students in computer science are the same : how important a place should be given to the computer ? What means should be deployed ? How should the mathematics curriculum be affected ?

Every opportunity for an elaborate reflection on these questions must be developped, because the decisions taken partly condition the future of the university teaching of fundamental subjects. The departments of mathematics of the universities of Leeds, Orsay and Rome have maintained research relations for quite a long time ; in 1983, they decided to extend their cooperation to the use of computer technology in the first cycle and they asked for financial help from the E.C.C. through the Office of Cooperation in Education. We first describe the framework and the development of the exchanges since 1983 and then we analyse their scope.

10.1 FRAME OF THE COOPERATION
The E.C.C. grants two kinds of financial aid for cooperation in university education :
"SHORT STUDY VISITS"
"The aim of this programme is to allow university teaching staff to enlarge their knowledge and their experience of the teaching in the other member states and to develop the possibilities of cooperation in the long run".

"COMMON STUDIES PROGRAMME"
"These aids are intended to encourage the cooperation between universities aiming at the conclusion of arrangements according to which :

i) The students spend a recognized and integrated part of their curriculum in another member state and/or

ii) Some parts of courses are given in each institute by teachers of at least one institute of another member states and/or

iii) The courses or parts of courses are jointly organized with an eye to their insertion in the curricula of all the participating institutes, without necessarily implying the mobility of teachers or students."

10.2 AID GRANTED AND EXCHANGES
1983-1984 : agreement n° SSV-83-243-F for *"short study visits"* (2.200 ecus). This aid has allowed financing of the travel expenses of 5 mathematics

teachers from Orsay to Leeds or to Rome. A publication (financed by the first cycle administrative unit at Orsay) has enabled information obtained during these visits to be diffused [9].

1984-1985 : contract n° SSV-84-367-F for the continuation of the "*short study visits*" (2.500 ecus). This contract covered the travel expenses of 5 mathematics teachers from Orsay :
- In England, to the universities of Leeds and Cambridge.
- In Italy, to the university of Rome and the Polytechnic Institute of Turin.

These visits have consolidated precious contacts and established new contacts with centers which have a large experience in the use of computers in mathematics teaching. An account of these visits, containing abundant didactic documentation, has been published by the first cycle administrative unit at Orsay [10].

The E.C.C. financed, in 1984-85, a project parallel to that of Orsay for the universities of Leeds and Rome ; in France the programme of visits concerned Orsay and Grenoble.

1985-1986 : contract n° JSP-85-447-F for the start of a "*common studies programme*" (2.500 ecus for teachers and 3.000 for students exchanges). The aid enabled a teacher from Leeds to come to Orsay (6 weeks), 8 students to do a visit (2 or 3 weeks) outside their university and financed a 3-day meeting in Orsay for teachers from Leeds and Rome to evaluate the exchanges.

1986-1987 : contract n° JSP-86-447-F for renewal of a "*common studies programme*" (2.500 ecus for the teachers and 2.000 for the students). This contract enabled a French teacher to stay in Leeds for 6 weeks, 6 students to spend 2 weeks outside their university, and 4 teachers (2 English and 2 French) to go to Rome to participate in a meeting to evaluate the results of the common studies programme and also to participate in ECM/87 (International Conference on the use of computers for the teaching of mathematics).

10.3. FORM AND CONTENTS OF THE EXCHANGES
The cooperation between teachers has developped in two ways :
- short visits or meetings (3 to 6 days), which allowed first a better knowledge of the teaching organization and of the computers available in each of our universities to be gained and an evaluation of the exchanges, a study of the prospects for this cooperation and progress in a rapidly changing teaching field to be carried out.
- teachers exchanges (4 to 6 weeks), which enabled active participation (during 6 or 9 hours a week) of the teachers in the mathematics teaching on computers in the first year of the host university. This experiment "*in the field*" allows a better appreciation of the differences and the analogies between European university systems : level of the students, coaching rate for the computer activities, teachers participation, didactic material, interaction with mathematics lectures, and so on [11].
The student exchanges have taken the form of visits of 2 or 3 weeks in the host university, sometimes followed by personal work leading to a written report. This work is marked and the mark obtained is used as a probationary period mark or an option mark, whenever the normal curriculum of the students offers this opportunity. The visit includes 4 or 8 hours of work on computers daily, an attendance in mathematics lectures and meetings with students at the host university.
Work on computers has two purposes : on the one hand the students can get accustomed to hardware or a language, which is different from those used in their university, and on the other hand they can develop a mathematics project

which is compatible with their level of studies. Up to now the visits have been supervised voluntarily by the common studies programme participants, and the computers have been supplied by the host university. The aid granted by the E.C.C. has covered travel and living expenses for the students. In 1987, the Leeds School of mathematics has financed the travel expenses of one student.

As an example, here are some themes proposed to the students.

Visit of three French students to Leeds : write an iterative program to solve Laplace equation in dimension two, with boundary conditions, display a graphical representation of the results ; program the solution of the unidimensional heat equation. The students have used the terminals of a mainframe (AMDAHL 470), first because the numerical techniques they used converged slowly, then because it provided them the experience of the handling of a big system.

Visit of two Italian students to Orsay in 1987 : part one, study the polynomial approximation of a continuous function on an interval, use Lagrange polunomials for approximate calculations of limits, derivatives and integrals (in PASCAL) ; part two, introduction to PROLOG and application to formal arithmetic.

10.4 EVALUATION OF THE COOPERATION

The reports of the teachers and the students who have benefited by exchanges are very favourable and demonstrate the value of this cooperation, which has enabled each of them to gain deeper scientific, technical or pedagogical knowledge and to participate actively in the university teaching of another European country.

The meetings and the exchanges of teachers carried out within this framework have been an invaluable way to elaborate common pedagogical documents, to evaluate the available didactic material in each of our centers and to carry out a reflection on the impact of computers on our teaching. Exchanges of medium length (from 4 to 6 weeks) prove the most fruitful and it seems to us that they should be continued and possibly extended to other European university centers.

The E.C.C. wishes the exchanges of students to be longer than one month and to cover a study period that can be recognized by their university. But university structures in the first cycle are very different in England, in France and in Italy. It is not possible, for the moment, to replace a term or a semester of studies in the first cycle in one of our universities by an equivalent period in another country, without a break in the normal curriculum. On the other hand, the aid granted by the E.C.C. remains limited, and an increase in the number of students exchanges or in their length would lead to tutoring problems in the host universities and require specific organization (travel, housing and so on). It seems to us that student exchanges in mathematics in the first cycle can only be experimental for the present time. But we must not underestimate their role in European integration and in preparing for future longer term exchanges.

11. CONCLUSION

For the interactions between mathematics and computers, we are at a stage where many questions exist and where the conclusions are very much related to the context of each university system. Let us sum up some reflections particular to Orsay.

All the departments accept and wish that the initial training in computer science for first cycle students should be undertaken by teachers of the fundamental subjects, with a view to clarifying and refreshing their teaching.

The technical services recognize the need for computer equipment for teaching in the first cycle and the requests formulated are satisfied, whenever general university policy permets.

In mathematics, can we assert that most teachers feel it is necessary to introduce the computer into their teaching and are ready to invest in it ? I think not. On the one hand the Orsay mathematics department is strongly oriented towards research in pure mathematics, and most teachers have not, at the present time, followed curricula where computers were used : either their research preoccupation does not leave them spare time to invest in computers, or they do not believe in the contribution of computers to the teaching of basic mathematics, or they think it is a marginal aid which concerns only a few enthusiasts. Thus the job of supervising the options at present and of reflecting on their impact on the contents of the first cycle syllabus is undertaken by less than 10 % of all the teachers.

Comparison with the School of mathematics of Leeds, where the number of teachers who belong to applied mathematics groups is higher than the in Orsay, shows greater teacher participation in teaching on computers [11].

Have the contents of the syllabus evolved under the influence of computers? The answer to the above question shows that in Orsay there has been no elaborate discussion on this subject. On the other hand, our first cycle mathematics courses are implanted in many other curricula, within the university (other subjects, *"Instituts Universitaires de Technologie"*) and outside the university (*"classes préparatoires"*, institutes for engineering studies) and it would be very difficult to make substantial changes in the different curricula if they were not synchronized. Nevertheless some trands are asserting themselves in the books that have been published in the last few years, at the level of secondary studies or the first cycle : more importance given to the numerical approach and to approximation methods ; in linear algebra, use of the Gauss method for the solution of linear systems, rather than the determinants method which requires heavier computation.

Are the students who have attended the proposed computer science options better trained in mathematics ? The opinion of the teachers who are responsible for these options is mixed. The options permit the students who have a high initial level to test, to consolidate and to widen their knowledge, by real experimentation with mathematical concepts : it brings a new dimension to their approach to the subject. For intermediate or low level students, the option is like an additional subject : either it has no influence on their understanding of mathematical concepts, or the computer aspect becomes the student's favourite subject to the detriment of the theoretical lectures.

Does the introduction of computers change the context of teaching ? The answer to this question is definitely positive, as in mentioned in many articles ([2] [3]), which perhaps gives an additional explanation to the reticence of some teachers about this introduction. The relations between professors and students in a computer laboratory are quite different from the relations in a classroom : the teacher can be far more often confronted with unexpected questions, and, in some cases, with situations when the student can be more familiar with the computer than he is himself. This situation can be uncomfortable for some mathematicians, but it allows others to improve their traditional mathematics teaching thanks to the new type of relation established during the practical sessions and to create a better pedagogical context. On the other hand, Orsay has little experience of other forms of computer use : illustration on a big screen, during lectures, of results, interaction between the students and the machine, either for personal mathematical experimentation, or within computer assisted learning.

Is European cooperation on the use of computers a major European issue ? Here again the answer is affirmative. On the one hand university teaching is a

European issue in itself, as the efforts of the E.C.C. towards teachers and students mobility show (ERASMUS project). On the other hand the mutual discussion of experiments carried out in our different university centers undeniably allows us to make progress in thinking about mathematics teaching and contributes to overcoming local reticence about the priority which the adaptation of our teaching to the world outside and to scientific progress represents.

REFERENCES

[1] The Influence of Computers and Informatics on Mathematics and its
     Teaching. An ICMI discussion document. L'Enseignement Mathématique 30
     (1984), 159-172.
[2] The Influence of Computers and Informatics on Mathematics and its
     Teaching. Proceedings from the Strasbourg symposium (1985).(Eds.) A. G.
     Howson and J.-P. Kahane, ICMI Study Serie,(Cambridge Univ. Press,1986).
[3] The Influence of Computers and Informatics on Mathematics and its
     Teaching. Strasbourg March 25-30, 1985, ICMI Supporting papers, IREM de
     Strasbourg(1985).
[4] Informatics and the Teaching of Mathematics in Developing Countries.
     Internation Symposium ICOMIDC-IFIP, Monastir (Tunisie), February 3-7,
     1986.
[5] Kahane, J.-P. Enseignement mathématiques, ordinateurs et calculettes.
     CIM Berkeley,1986.
[6] Burkhardt, H. Computer-aware curricula : Ideas and realizations, in
     [2], p. 147-155.
[7] Goetgheluck, P. Documents de travail sur ordinateur. Univ. de Paris-
     Sud, Math. Bât. 425, 91405 Orsay Cedex, 1987.
[8] Cottet-Emard, F. Module M2A, Mathématiques et Informatique. Univ.
     Paris-Sud, Math. Bât. 425, 91405 Orsay Cedex, 1987.
[9] Déchamps, M., Emery, F., Garcia, F., Goetgheluck, P. L'enseignement des
     mathématiques et l'utilisation des micro-ordinateurs en premier cycle
     universitaire dans les universités de Leeds, Paris-Sud et de Rome.
     Visites d'étude de courte durée financées par la C.C.E. Univ. de Paris-
     Sud, UER 1er cycle, 91405 Orsay Cedex, 1984.
[10] Cottet-Emard, F., Garcia, F., Rivier, M. L'enseignement des
     mathématiques par les moyens informatiques en premier cycle
     universitaire. Rapport des visites de courte durée financées par la
     C.C.E., Univ. Paris-Sud, UER 1er cycle, 91405 Orsay, 1985.
[11] Goetgheluck, P. Rapport sur un séjour à l'Ecole Mathématiques de
     l'Université de Leeds en 1986. Univ. de Paris-Sud, Math. Bât. 425, 91405
     Orsay, 1987.

ECM/87 - Educational Computing in Mathematics
T.F. Banchoff et al. (editors)
© Elsevier Science Publishers B.V. (North-Holland), 1988

STUDY OF SEQUENCES USING A PERSONAL COMPUTER

Pierre JARRAUD et Christine LAURENT

Université Pierre et Marie Curie Mathématiques 45-46 5e étage
4, place Jussieu 75252 PARIS-CEDEX 05 FRANCE

We present here two examples of using educational computing in teaching the sequences of reals. The first one uses a software giving a graphic representation of the convergence of sequences, the second one is given by the use of a spread-sheet (Multiplan by Microsoft) for numerical study of real sequences.

A. GRAPHIC REPRESENTATION OF THE LIMIT OF REAL SEQUENCES

The aim of this software, written in BASIC and running on IBM-PC-compatible personal computers with graphic card, is to give to the students in the first year of university a more concrete image of the notion of convergence of a sequence of real numbers.

1. PRESENTATION OF THE SOFTWARE

This software enables a graphic representation of real sequences with general term of the following type:

$$U(n) = f(n, U(n-1), U(n-2))$$

The representation is given by little crosses at the points of coordinates $(n, U(n))$ for n between 0 and T, T being chosen by the student:

-if T < 64 all points appear on the screen,

-if T ≥ 64 only points such that $\dfrac{n}{[T/64+1]}$ is an integer appear on the screen.

The scale on the Oy axis is also chosen by the student to obtain the best scaling of the graph of the sequence or a good precision round a chosen value (the limit for example).

Moreover for selected values of L and EPS > 0 (L being the limit of the sequence when it converges) the lines of equation y = L-EPS and y = L+EPS are drawn on the screen.

The integer N such that $|u(n)-L| < EPS$ as soon as N < n ≤ T and the line x = N are displayed. The following feature appears: all the crosses on the right of the line x = N are between the lines y = L-EPS and y = L+EPS.

2. THE TUTORIAL SESSION

We supply a list of sequences, the definition of which is compatible with the software and such that a first year student can find whether they are convergent and what their limit is.

2.1 We ask for a graphic representation of the sequences and a classification in (apparently) converging and non-converging sequences. At this stage L and Eps are nul and the lines y = L- EPS and y = L + EPS are equal to the Ox axis. Results are stored in an array in view of subsequent theoretical proof.

2.2 For each *apparently* convergent sequence of 2.1 students calculate the limit and run the software, L set to the calculated value and EPS successively 0.1, 0.05 and 0.01 determining three integers N_1, N_2 and N_3. Students are then asked to analyze the results obtained: formalization of the notion of convergence, study of the speed of convergence...

Remarks:

-if a student has failed in his calculation of the limit or in the determination of the nature of the sequence, he will spot his mistake: even for big T he will not get the values of N_1, N_2 and N_3 given by the software.

-if EPS = 0.01 it is often necessary to change the scale of Oy to display the band L - EPS < y < L + EPS and to choose T big enough to get N.

-if T ≥ 64 not all the points are displayed on the screen and the student may have a misleading idea of the behaviour of the sequence: for example if $U(n) = (-1)^n$ and 64 ≤ T < 128 only points for even n are displayed and the sequence is apparently constant. The calculation of N takes all values of U(n) for n < T.

-At the end of the session the students are required to give their opinion about the computer as a tool for studying real numbers sequences. It is worthwhile to insist upon the computer limits for calculation and upon the risk of grossly erroneous results due to truncation errors. The example of the sequence $U(n) = n!/k^n$ for big values of k may draw the students' attention on the necessity of proofs and on the risks of too much trusting the machine during the study of sequences.

3. GRAPHS

The following pages are examples of graphs obtained with this program.

We chose the following sequences:

(1) $U(n) = \cos(n \frac{\pi}{6})$

We took T = 50 and Y between -1.5 and 1.5.

(2) $U(n) = \dfrac{n^2-25}{2n^2+1}$ (limit L = 1/2).

We took T = 50, EPS = 0.05 and Y between -0.5 and 0.7 on the Oy axis.

(3) $U(n) = \sin\left[\dfrac{1}{\sqrt{n+1}}\right]$ (limit L = 0).

We took EPS = 0.01, T = 15000 and Y between -0.05 and 0.1.

(4) $U(n) = n!/k^n$ with k = 20.

On the first graph T = 20, EPS = 0.005 and Y between -0.01 and 0.01. On this picture the sequence seems to converge towards 0. In fact this sequence tends to infinity with n and a better representation of its behaviour is given by the second graph where T = 50 (the scale being the same on the Oy axis).

With such an example one can point out to the students, that it is not possible to decide whether a sequence converges or not just by the calculation of the first terms.

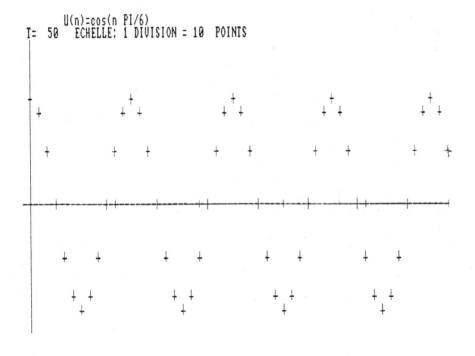

U(n)=cos(n PI/6)
T= 50   ECHELLE: 1 DIVISION = 10 POINTS

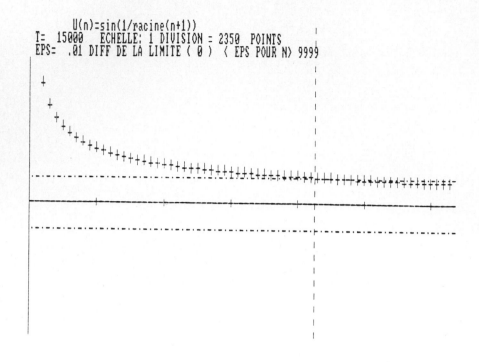

```
           U(n)=sin(1/racine(n+1))
   T=  15000   ECHELLE: 1 DIVISION = 2350  POINTS
   EPS=  .01 DIFF DE LA LIMITE ( 0 ) < EPS POUR N> 9999
```

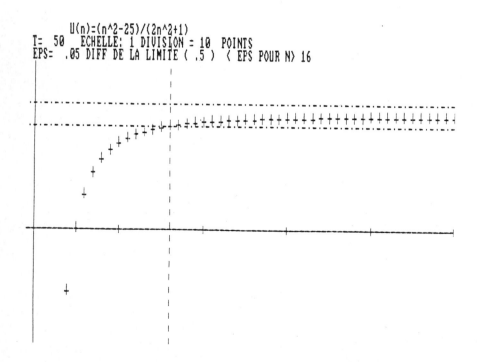

```
         U(n)=(n^2-25)/(2n^2+1)
   T= 50   ECHELLE: 1 DIVISION = 10  POINTS
   EPS= .05 DIFF DE LA LIMITE ( .5 ) < EPS POUR N> 16
```

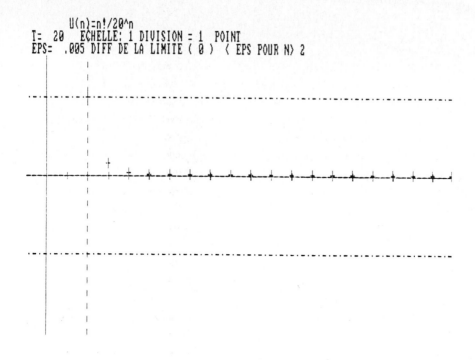

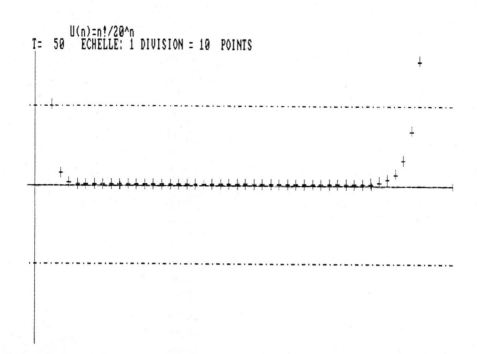

B-MULTIPLAN

1. WHY A SPREAD-SHEET?

Our purpose was to find a very simple-to-use process, without any previous knowledge of computer programming, enabling a numerical study of real sequences. Thus the actual programming (by the students) in a language such as Pascal or Basic was excluded and due to lack of time we could not carry out ourselves a ready-to-use software.

We chose to use a **spread-sheet,** in our case **MULTIPLAN** by **MICROSOFT** available in our computers room.

Using a spread-sheet (and Multiplan in particular) presents a lot of advantages:

-it is easy to get numbers arrays, and there is no need of tabulation's instructions.

-the calculation's precision (14 digits) is good and it is easy to modify the number of displayed digits by modifying the width of the column.

-it is easy to modify parameters or initial values. Moreover one can enter expressions like "square root(2)" (try to do so in Basic or Pascal!..).

-the structure of the sheet is convenient for the study of real sequences as most frequently met $(u_n = f(n, u_{n-1}, u_{n-2}))$ and the notions of variables and relative coordinates are clearly emphasized.

-the possibility to save and restore "sheets" enables a track of already made calculations to be kept. In addition, by supplying the students with a "fair copy" it prevents them for being stucked in machine related difficulties (even in this case the parameters modifications remain free).

2. WHAT IS POSSIBLE?

We showed to the students the interaction between computer calculation and mathematical thinking.

Let us consider the following example:

EXERCICE 1: Approximation of $\sqrt{2}$ by rational numbers.

A)Consider the numbers $\theta = 1 + \sqrt{2}$ and $\theta' = 1 - \sqrt{2}$. Define $u_n = \sqrt{2}\,\theta^n - \sqrt{2}\,\theta'^n$ and $v_n = \theta^n + \theta'^n$.

1)Show that u_n and v_n are even integers, that u_n and v_n grow to infinity when n tends to infinity, and that $\left| \sqrt{2} - \dfrac{u_n}{v_n} \right| < \dfrac{4}{v_n^2}$ .Deduce that u_n / v_n is a rational number giving a very good approximation of $\sqrt{2}$.

2)Find a polynomial of degree 2 with roots θ et θ' and prove that u_n and v_n satisfy the relationships:

$$\begin{cases} u_{n+2} = 2\,u_{n+1} + u_n \\ v_{n+2} = 2\,v_{n+1} + v_n \end{cases}$$

Show that $u_0 = 0$, $u_1 = 4$, $v_0 = 2$, $v_1 = 2$.

Using the spreadsheet, calculate the first 40 values of u_n, v_n, u_n/v_n, $4/v_n^2$ *and deduce from this calculation a rational number approximating* $\sqrt{2}$ *within* 10^{-27}.

The students are supposed to get the following array:

Approximation de racine(2) par des rationnels
****

n	Un	Vn	Un/Vn	4/Vn^2
1	0	2	0	1
2	4	2	2	1
3	8	6	1,33333333333	0,1111111
4	20	14	1,42857142857	0,0204082
5	48	34	1,41176470588	0,0034602
6	116	82	1,41463414634	0,0005949
7	280	198	1,41414141414	0,000102
8	676	478	1,41422594142	1,751E-05
9	1632	1154	1,41421143847	3,004E-06
10	3940	2786	1,41421392678	5,153E-07
11	9512	6726	1,41421349985	8,842E-08
12	22964	16238	1,1142135731	1,517E-08
13	55440	39202	1,41421356053	2,603E-09
14	133844	94642	1,41421356269	4,466E-10
15	323128	228486	1,41421356232	7,662E-11
16	780100	551614	1,41421356238	1,315E-11
17	1883328	1331714	1,41421356237	2,255E-12
18	4546756	3215042	1,41421356237	3,87E-13
19	10976840	7761798	1,41421356237	6,639E-14
20	26500436	18738638	1,41421356237	1,139E-14
21	63977712	45239074	1,41421356237	1,954E-15
22	154455860	109216786	1,41421356237	3,353E-16
23	372889432	263672646	1,41421356232	5,753E-17
24	900234724	636562078	1,41421356237	9,871E-18
25	2173358880	1536796802	1,41421356237	1,694E-18
26	5246952484	3710155682	1,41421356237	2,906E-19
27	12667263848	8957108166	1,41421356237	4,986E-20
28	30581480180	21624372014	1,41421356237	8,554E-21
29	73830224208	52205852194	1,41421356237	1,468E-21
30	178241928596	126036076402	1,41421356237	2,518E-22
31	430314081400	304278004998	1,41421356237	4,32E-23
32	1038870091396	734592086398	1,41421356237	7,413E-24
33	2508054264192	1773462177794	1,41421356237	1,272E-24
34	6054978619780	4281516441986	1,41421356237	2,182E-25
35	14618011503752	10336495061766	1,41421356237	3,744E-26
36	35291001627284	24954506565518	1,41421356237	6,423E-27
37	85200014758320	60245508192802	1,41421356237	1,102E-27
38	205691031143920	145445522951120	1,41421356237	1,891E-28
39	496582077046160	351136554095040	1,41421356237	3,244E-29
40	1198855185236200	847718631141200	1,41421356237	5,566E-30

It is easy for the students to get it, but it is more difficult for them to get the rational number close to $\sqrt{2}$ within 10^{-27} : u_n/v_n is displayed on the screen with 14 significant digits and many students stay perplex. And the teacher needs to remind them the *theoretical* part to show them that u_{38}/v_{38} is the right answer.

We used also Multiplan for illustrating theoretical exercises: Newton's method, arithmetic-geometric mean and calculation of elliptic integrals.

ECM/87 - Educational Computing in Mathematics
T.F. Banchoff et al. (editors)
© Elsevier Science Publishers B.V. (North-Holland), 1988

MATHEMATICS IN A COMPUTER AGE

Hervé Lehning

Ateliers Logiciels de l'Enseignement Supérieur
(ALE Sup)
13 rue Letellier 75015 Paris,   France

There are many ways of using computers in Mathematics :
- *for students programming,*
- *as calculation tools,*
- *as problem solving assistants.*
Those three ways are important and are still being studied in France, particularly by ALE Sup which coordinate software workshops at University level. In this paper, I would like to develop these three ways and their consequences on the teaching and the learning of Mathematics.

1. STUDENTS PROGRAMMING.

1.1. Why ?

It is the use one thinks of and as a matter of fact, it is a very formative activity. But *why* particularly in Mathematics ? We think there are three reasons for this :

- using black boxes without ever looking inside seems inadequate to the training of scientists,

- coding an algorithm allows a better comprehension of it,

- it is a good activity to learn consistency which is very important in Mathematics.

1.2. Dangers.

But there are several dangers :

- programming is *time consuming,* this is the first and the most important danger,

- the second one is that many problems are not Mathematical, for example the input / output ones.

1.3. A solution : The Modulog system.

(i) Presentation.

ALE Sup has a solution to those problems : it is a system of software components, called *Modulog*, which greatly simplifies the programming (in Turbo Pascal) of small scientific calculations by students. This library is divided in two parts :

- the basic kernel which allows one to manage the inputs and outputs;

in particular, it contains a graphic kernel admitting all usual
cards and a component of the input of numbers, functions and
matrices (in numerous and easy ways).
- components which allow one to do a lot of usual computations.
The detailed description of this system (which is in use in
France : 2000 copies have been sold in 1987) is available on
request.

(ii) An example.

Here, I shall give a very simple example of the use of the
basic kernel. With some files, we can use some types, functions
and procedures, more particularly these :

Types : Numerical_function, Matrix and Vector,

Procedures :

Input_real which allows one to input real in a symbolic way (for
example (1 + sqrt(5))/2),

Input_function which allows one to input function in the same way
(for example (1 + sin(x))*exp(-x)),

Input_matrix which allows one to input matrix in a nice way and
also output_matrix, input_vector, output_vector.

Window which defines the screen as a window on the plane, move and
draw.

Functions : Evalue1, 2, 3 or 4 to compute a function input through
the procedure Input_function with the recovery of overflow error.

- Listing.

For example, here is the listing of a curve sketcher on an IBM
PC compatible :

```
Program Curve ;
($i MATH.LIB )
($i ENTREES.LIB )
($i G640X200.CGA )
( if you use a CGA card, we also have the following files :
G320X200.CGA, G720X348.HER, G640X350.EGA, G640X400.OLI )
($i GRAPHE.LIB )              ( those included files allow us to use the
objects defined in them )
Var    x,y                    : Numerical_function      ;
       x_min,x_max            : Real                    ;
       y_min,y_max            : Real                    ;
       t,t_min,t_max,step     : Real                    ;
Begin
  write('x(t) = ')  ; Input_function('t',x)  ;
  write('y(t) = ')  ; Input_function('t',y)  ;
```

```
   write('t_min = ') ; Input_real(t_min)    ;
   write('t_max = ') ; Input_real(t_max)    ;
   write('step = ')  ; Input_real(step)     ;
   write('x_min = ') ; Input_real(x_min)    ;
   write('x_max = ') ; Input_real(x_max)    ;
   write('y_min = ') ; Input_real(y_min)    ;
   write('y_max = ') ; Input_real(y_max)    ;
   InitGraphix ;   ( for graphix initialization )
Window(x_min,x_max,y_min,y_max) ;
   t := t_min ; Move(Evalue1(x,t),Evalue1(y,t) ;
   repeat
     t := t + step ;
     draw(Evalue1(x,t),Evalue1(y,t)) ;
   until t > t_max ;
End.
```

   - **A run.**

   Figure 1 is a copy of the screen after a run time :

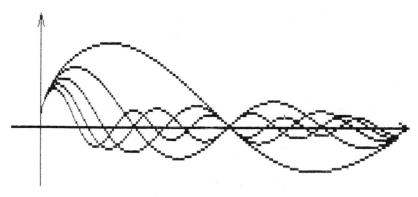

FIGURE 1

On this figure, we can see that the sequence of functions defined by : $f_n(x) = \dfrac{\sin nx}{n \sqrt{x}}$ seems to converge uniformly. Then, it is easier to prove : see paragraph 3 for the study of this kind of attitude.

2. A CALCULATION TOOL.

Using computers as calculation tools is not a new idea and scientists have used it for a long time but the software packages in use are new :

   professional ones are too complicated and not pedagogic. When we have good software, we can wonder which influence it has on teaching and learning Mathematics.

2.1. The notion of elementary object is evolving.

(i) generalities.

The idea that it is possible to work with complicated objects as with real numbers is fundamental in mathematics, nevertheless it remains abstract for many students. The possibility of doing quickly a great number of these calculations gives a more concrete meaning to these objects; in some way, they become primary. For example, linear algebra becomes more concrete through the use of a software allowing to compute quickly on matrices.

(ii) an example.

On the *Calcul matriciel* software (see [4]), we input the matrix as with a spreadsheet and then we can achieve any usual calculation. For example, it is very easy to compute eigenvalues with the power method. We input the matrix A, the functions :
$$f(v) = A * \frac{v}{\| v \|} \quad \text{and} \quad g(v) = \frac{\langle v \diagup A*v \rangle}{\langle v \diagup v \rangle}.$$
The n-th iterate of f is defined in such a way : $\phi(i,1,n,f(v))$ (the same as Σ, see 3.2.). So, we quickly obtain an approximation of the dominant eigenvalue which was impossible in the old days. And so, this domain becomes more concrete for students. A good proof of this is that they are now able to discover some properties on matrices without any help, for example the absolute value of the above sequence is increasing if A is symmetric. Also remark that it becomes easy to give a constructive proof of the theorem upon such matrices.

2.2. The notion of result is changed.

For example, what is the meaning of such a question ?
More exactly, what is the expected answer to this initial value problem : $(x^2 - 1) y' - x^2 y = 1 \qquad y(o) = o$?
Is it the formula :

$$y = \sqrt{\frac{1 - x}{1 + x}} \; e^x \int_o^x \frac{e^{-t}}{(t^2 - 1) \sqrt{\dfrac{1 - t}{1 + t}}} \, dt$$

true for x in]-1 , 1 [, or the formula : $\quad y = \sum_{n = o}^{\infty} a_n x^n$

$a_o = o \quad a_1 = -1 \quad a_2 = o \quad$ and $\quad a_{n+1} = \frac{1}{n + 1} [(n - 1) a_{n-1} - a_{n-2}]$
or the use of a numerical method like Runge-Kutta with a qualitative study (monotony, limit) which does not require an explicit formula ? Proving those properties is possible without any formula : using the uniqueness theorem and the equation is sufficient. Also see paragraph 3.4.

2.3 The notion of proof is changed too.

Everybody knows the story of the demonstration of the four colour theorem but it is sometimes wrongly thought that this type of reasoning cannot be found in undergraduate student work. It is not true, the most simple examples are in Number Theory :

for example, find the natural numbers equal to the sum of the cube of their digits in the decimal system.

It is very easy to prove that such a number has less than four digits; then with a computer it is possible to check all those numbers. It is also possible to find the solutions by demonstrating some properties of the solution. Does one of those prove better ?

2.4. Therefore, the curriculum will change.

For example, we have to answer some questions like :

What is   usual function, what is a formula, what is a proof ? And we have also some partial answers :

In an answer to a problem, we have to distinguish qualitative and quantitative aspects better (see paragraph 3.4).

The use of computers requires more abstract mathematics; as a matter of fact, a powerful tool is always more difficult to use. But computers allow a better understanding of mathematical phenomena because testing on mathematical object becomes easy :

computers are Mathematical Problem Solving Assistants.

3. A PROBLEM SOLVING ASSISTANT.

The third use of a computer is in fact nearer its use in usual works : it is an assistant to solve the problem itself. History shows the importance of an experimental attitude in the process of discovery in Mathematics. It is to be noticed that *"modern"* style in teaching has not suppressed this type of methods used by mathematicians.

Nowadays, the use of computers gives more value to this attitude thanks to the power of calculation we get this way. This type of attitude has been experienced in France, there are two examples in paragraph 1.3 (study of a sequence of functions) and in paragraph 2.1 (use of power method), here are four examples which use software packages developped in one of our Workshops (the Ecole Centrale de Paris one , also see [1] and [2]) :

3.1. A classical exercise of calculus.

Find the principal part near 0 of :

$$f(x) = \tan(\sin(x)) - \sin(\tan(x)).$$

This exercise is a trap because the order of this function is 7 ! But with a little testing, it is very easy to find the right order directly : The limit of $\dfrac{\ln |\frac{f(kx)}{f(x)}|}{\ln k}$ is the order.

We also compute the value of this function for k = 1.1 and x = 0.1 and we have a good estimation of the order to be used. When we have this order, the computation is easier to achieve.

3.2. Display of Gibbs' phenomenon.

On the *Etudes Graphiques* software package, the partial sums of the Fourier series which sum is the square wave function are defined in such a way : f(x,n) = 4/pi$\sum$(i,1,n,sin((2i-1)x)/(2i-1)) which is the usual way of writing them in Mathematics.

Then for n = 20, we obtain figure 2. We already see the phenomenon on this visualization, figure 3 represents the new window which we have decided to view in order to examine the phenomenon better.

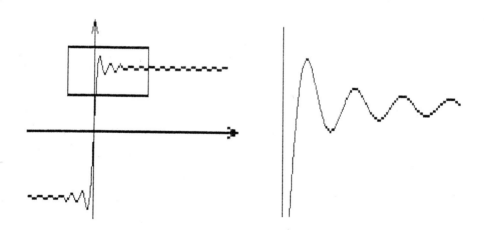

FIGURE 2   FIGURE 3

Then we get the phenomenon (on an experimental point of view); It is then easy to check it on the theoretical point of view.

3.3. Study of a numerical series.

If the series of general term : $u_n = \left(\dfrac{1}{\ln n} \right)^{\ln n}$ is to be studied.

On the *Etudes des Suites* software package, it is defined in this way : Un = (1/ln(n))^ln(n). While calculations are being done, we

can notice a rather slow convergence to 0. This virtually suppresses a convergence of a geometrical type, which is yet the most current idea among students who are not experimented enough. The calculation of Un / Un-1 allows the checking of this hypothesis. In order to do so, we define a new sequence v_n, which is done this way : $v_n = \dfrac{u_n}{u_{n-1}}$. We obtain, by writing the terms 10 by 10 : 0.813 0.894 0.925 0.942 0.953 0.960 0.965 0.969 0.972 ... so the sequence seems to converge to 1. So it seems more probable that the series is of the Riemann type, then we calculate :
$w_n = -n \ln(v_n)$ and we obtain, still with the same step : 2.042 2.228 2.318 2.379 2.425 2.462 2.492 2.518 2.541 ... which leads us to think of an increase of the type $\dfrac{A}{n^2}$ (which insures the convergence of the series). So we define : $x_n = n^2 u_n$ and we obtain : 17.735 16.638 15.091 13.748 12.623 11.678 10.875 10.183 which seems to confirm the latest hypothesis. Once this conjecture completed, it is easy to demonstrate :
$\ln(x_n) = \ln n [2 - \ln(\ln n)]$ converges to $-\infty$ so x_n to o.

3.4. Study of a differential equation.

When we face the differential equation : $y' = x - 1/y$,   we can begin by sketching the family of integral curves (see figure 4) :

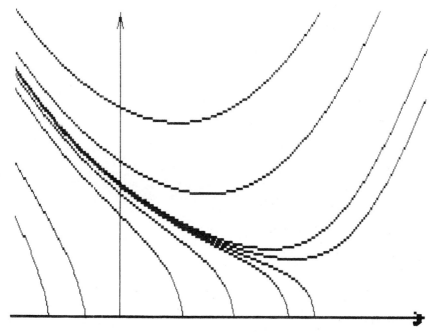

FIGURE 4

Then it seems that there is a solution y defined and decreasing on $\mathbb{R}$ which tends to $+\infty$ at $-\infty$ and to zero at $+\infty$ such that :

- the integral curves above y are defined on $\mathbb{R}$, decreasing until they meet the curve which equation is : $x\,y = 1$, and then increasing, tending to $+\infty$ at $-\infty$ and $+\infty$,

- the integral curves below y are defined and decreasing on $]-\infty,a[$ where a belongs to $\mathbb{R}$ and tend to zero at a.

Moreover, the sketching gives some ideas on a proof :

For each $x > 0$, there are two solutions y_x and z_x which are defined on $\mathbb{R}$ and $]-\infty,x[$ such that $y_x(x) = 1/x$ and $\lim_{t \to x} z_x(t) = 0$.

Let $f(x) = y_x(o)$ and $g(x) = z_x(o)$, it is easy to show that f is increasing and g decreasing and : $o < f(x) - g(x) < 1/x$. So f and g have a common limit l and the solution y such that $y(o) = l$ is the function we are looking for.

In this case, we see a problem which seemed very difficult in the days before the appearance of computers (also see [2] and [3]).

4. CONCLUSION.

Through these examples I developped, we can see that it is not only the centres of interests which are evolving because of computers (algorithmic, constructive solutions) but also the way of doing mathematics. Essentially, computers are assistants for problem solving through checking and testing. This idea is close to Seymour Papert's (see [6]).

REFERENCES.

[1] Hervé Lehning and Daniel Jakubowicz, Mathématiques par l'informatique individuelle(Masson, Paris, 1981).
La matemica e il personal computer (Italian).
Matemáticas para la informática personal (Spanish).
[2] Hervé Lehning, Mathématiques Supérieures et spéciales(Masson, Paris, 1985).
[3] Michèle Artigue and Véronique Gautheron, Systèmes différentiels étude graphique(Cedic, Paris, 1983).
[4] Atelier logiciel de l'Ecole Centrale, Etudes graphiques, Etudes des suites et séries numériques, Calcul matriciel, Etude des surfaces, 1986-87. See presentation of software.
[5] ALE Sup, Modulog, 1987. See presentation of software.
[6] Seymour Papert, Mindstroms, children, computers, and powerful ideas(Basic Books, New York, 1980).

ECM/87 - Educational Computing in Mathematics
T.F. Banchoff et al. (editors)
© Elsevier Science Publishers B.V. (North-Holland), 1988

Computers and the Teaching of Statistical Practice

E. J. Redfern

Department of Statistics
University of Leeds
Leeds LS2 9JT, England

We consider how the speed and range of calculation available on a computer can influence the way in which statistics is taught. In particular we discuss the continuing role of practical work, the impact of statistical packages and a possible role for expert systems.

1. INTRODUCTION

Since Yates [13] described the arrival of the computer as the beginning of the second revolution in Statistics, (the first occurred with the arrival of the calculator), its impact on both research and the practice of statistics has been extensive.

Computer intensive methods, such as those described by Diaconis and Effron [6], use methods of analysis and interpretation based on the use of computation rather than mathematical theory. Robust methods (Huber [9] are almost as easy to use as the classical methods based on normal distribution theory. Exploratory data techniques (Tukey [11]) and diagnostic aids to model analysis (Cooke [5], Weisberg [12] and Atkinson [1]) are areas which are greatly enhanced by the increased speed of computation and the graphical facilities available using a computer. Graphical methods in particular are much easier to put into practice and statistical packages are now widely available on computers of all sizes.

One aspect of these is a decline of importance of Normal distribution theory and the need to use statistical measures whose properties can be developed analytically. Several models can be fitted and compared and questions such as

1) how good is the fit?

2) are there any bad points? and

3) how can the model be improved?

can be more easily tackled. Speed [10] makes a case for the teaching of

Statistics at University to begin to reflect these changes in the nature of the subject.

Statistical practice has benefited from this freedom from earlier computational constraints, allowing much more detailed and complex methods of analysis to be used.   However, one effect of this is the ease with which sophisticated tools can be used by the non-expert, sometimes with disastrous consequences.   Consequently a greater responsibility rests on people who teach statistics, particularly those who teach non-mathematicians.   Students need to be taught to appreciate not only the potential applications of statistical models or methods but also the limitations, particularly of the assumptions on which they are based.   The fact that it is easier to turn the handle makes the need for awareness of the outcome much greater.

Data analysis and practical work are a traditional part of statistical education and for many years the calculations were performed on mechanical, electro-mechanical and more recently on electronic calculators.   While these facilitated slightly more extensive calculations, the techniques used were broadly those used for hand calculation.   Access to computers has resulted in a dramatic change in the type and the amount of computation possible.   By using statistical packages we can teach a wider range of data summary and exploratory techniques, and expect students to consider alternative models and their validity.

Furthermore improvements in micro-computers, particularly the graphic capabilities, have resulted in their use for illustrating concepts.   The latest generation of micro-computers, being faster and having more memory, should see a merging of these two uses and overcome one of the current limitations of statistical packages - the lack of good interactive graphics.

Based on our experience at Leeds, we consider several aspects of the computer's influence on statistics teaching and the role that it should take. In particular we concentrate on the effect of the computer on teaching elementary concepts, the influence of statistical packages and teaching the wider aspects of statistical analysis.

2. TEACHING FOUNDATIONS

Using a computer should permit a greater appreciation of the foundations of statistical practice and the interpretation of data. It allows us to emphasise the choice of hypotheses, the interpretation of the results of tests, and the validation of the assumptions, where possible, as the important statistical features.   The derivation of the appropriate formula, following through from the assumptions, is not as crucial to understanding the application and interpretation.

The main objectives of a first statistics course are to develop an understanding of statistical concepts and ideas, elementary data analysis and the decision-making process.   The computer can assist us in several ways in the way we do this.   It also effects what is taught and the order in which it is taught.   To illustrate these points we consider two aspects of this influence, namely the illustration and understanding of concepts and the use of statistical tables.

2.1 Illustrating Concepts

While many students can cope with the development of the mathematics they all too often have difficulty in interpreting the results.   A verbal description is not often understood, particularly by the weaker students. As a result subsequent ideas built on this weak basis are not secure.   The graphical facilities of a computer together with the use of simulation can help to strengthen this foundation.

Holmes [8], among others, has shown how software using a micro-computer can be used to illustrate, clarify and reinforce concepts such as significance levels, confidence intervals and the relationship between distributions like the Binomial, Poisson and Normal.   For example, repeated simulation of confidence intervals for a parameter reinforces the idea that they do not always contain the true value and the rate at which this event can occur.   Software of this kind can be used to support the mathematical development of the subject by helping to clarify the interpretation of the mathematics.   However, we need to ensure students understand the principles of simulation and how the results relate to real data.

Practical experimentation is important in helping to achieve this and can be a useful interface between the computer and the theory.   Green [7], for example, has illustrated the difficulty 11-16 year olds have in understanding probability concepts.   Applying his tests to undergraduates at Leeds wo have found a similar lack of understanding even at the end of a traditional introductory statistics course for mathematicians.   It is particularly apparent that students do not fully understand the short term variability that occurs in data.   Their interpretation is based on the long term expectation derived by the mathematics from the assumptions made.

For example the standard definition of randomness relates to selecting items from a group in which each item has an equal chance of being chosen. This is an easy idea to imagine and seems fairly easy to understand.   For this reason it is usually presented without further explanation.   However, until we actually select a sequence of items from the group according to some random procedure, we fail to see how difficult it is to achieve or confirm. Anyone who has repeatedly tossed a coin, recording the results, will know how

easy it is for the process, even though we go to great extents to ensure that
it is random, to produce results that appear to favour one side at the
expense of the other.   To show that the procedure is in fact random we need
to gather extensive information.   We need to show that the actual
occurrences of events, in the long run, agree with the assumption that they
are equally likely to occur and also to establish that there is no system or
order in the way that they occur.

The related concept of independence is also hard for students to accept
while the gambler's fallacy of dependence is much more believable.
Presented with a short sequence of results generated by a random mechanism
such as tossing a coin even mathematics students let the evidence of a short
sequence dominate the prediction of the next outcome rather than basing that
prediction on the assumption that the coin is unbiased.   Thus when presented
with a sequence of four heads, a tail is usually stated as the more likely
outcome on the next throw.

Another idea fundamental to statistical thinking is that of variability.
Mathematical models and theory tend to emphasize the homogeneity and
long-term stability of data rather than the short-term variability.   It
should be asked whether students understand how variable real data can be -
even in the case of symmetric distributions.   We need to make the point that
data values from the same distribution are not alike and the extent to which
they vary among themselves is important hence the need to measure the extent
to which they are dispersed, spread out or bunched.

Variability is usually introduced when discussing measures of spread such
as the standard deviation.   If students are asked to describe standard
deviation the usual reply is that it is a measure of spread followed by a
quotation of the formula.   This reflects on the standard text book approach
and the way in which many of us teach it.   It is questionable how many
students really understand that it is a measure of the departure from central
tendency and that the location of the data has no influence. An approach
through the mean deviation as presented in some modern statistics texts would
appear to be much more informative but it still does not convey the idea of
short term variability that is present in data.

These concepts are fundamental to simulation, modelling and assessing the
quantities we use as estimators, yet many of us make use of computers to
perform simulations to illustrate concepts or compare models without ensuring
that the students clearly understand the concepts on which the simulations
and their interpretation are based.   Thus while a computer has a very
important role to play in the teaching of statistics, particularly as a
result of this ability to perform simulations, it is necessary to bridge the

gap between theory and simulation. This is one of the areas where practical experience gained by simple experimentation has such an important role to play both in developing and understanding concepts and helping to introduce the computer as a tool for further development of the ideas.

Historically the limited ability to perform calculations seriously restricted the amount of practical work that could be performed. Thus while practical experiments have often been considered they have not been developed to their full potential. Using a practical framework that is a mixture of experimentation and computation we can overcome this. The ability of the computer to perform simulations and large amounts of calculation can extend the scope of the experimentation. Its use should however be tempered with care since, when a simulation is run, it is easy to mask both what is really happening and the ideas behind the calculation.

A second problem is the time factor. It must be seen that there is benefit to be gained from performing the experiment in order to justify the time spent doing it. While it is a well-established practice in physics and chemistry that experiments are performed to develop and illustrate theoretical results, this is not the case with statistics. The subject is deeply rooted within the mathematics framework. We can, however, reap great benefit from also using a practical approach in which to establish and explore ideas. This can in turn generate interest in the subject and help communicate ideas to the less mathematically able students.

Teaching statistics can therefore usefully benefit from a mixture of experimentation and computation. The latter being used to carry out some calculations and perhaps more importantly, at the introductory level, to perform simulations which extend and further illustrate some of the ideas. Practicals should be kept short and interesting and not attempt, at least initially, to introduce too many new ideas at once. It is also useful to explore ideas in a variety of ways to assist in the building of a student's confidence in their understanding of statistics Once this confidence has been achieved they can be used to develop ideas related to modelling data reflecting the theoretical ideas that are being developed elsewhere in the course. Estimation and sampling distributions, for example, can best be introduced using a mixture of experimentation and computer simulation.

2.2 The role of Statistical tables

Use of statistical tables also form a necessary part of most traditional introductory courses; it is therefore useful to conclude this section by consideringthe affect of computers on this aspect of our teaching. Bousbaine and colleagues [3] even go so far as arguingthat they are obsolete since the values required can be very easily generated using a programmable

calculator or computer.   Most computer packages supply   p   values based on
the standard set of assumptions, as part of the output to procedures while
others such as   MINITAB, have simple procedures for calculating any necessary
values from a wide range of distributions.

The emphasis therefore is on learning how to interpret these in the
context of the problem, yet working with MINITAB we can still require the
student to know how to calculate them from a distribution.   The   p   value
reinforces the students idea of the costs involved in the decision rather
than always working with the 5% value.   The observed statistic compared with
the 5% value from the distribution is not so easily interpretable in terms of
the relative costs of making the wrong decision as the p value.

3. USING STATISTICAL PACKAGES

Statistical packages have had a significant effect on the amount of
statistical analysis that can and is being done.   They allow a wide range of
methods and models to be used very easily.   Furthermore, constant revision
of packages means that many of the more recent developments are available to
expert and non-expert alike.   They are also having an influence on the way
that statistics is taught, not least because of the type of information they
make available.   At Leeds our experience in using them as a teaching tool
has been concentrated on three packages, - MINITAB, SAS and GENSTAT.   We
illustrate the effect packages can have both on the way that we teach
statistics and the content of courses by considering a simple example.

3.1 Introductory Courses

MINITAB is mainly a teaching package which is simple to use and
facilitates most of the problems involved in introductory and non-specialist
courses.   Data is stored in columns, constants or matrices and is easy to
enter and use in the various statistical procedures available. A small
disadvantage is that data editing and manipulation is not always
straightforward although this situation has improved with version 5.1.

To illustrate the role and influence the package can have on our teaching
consider a two sample problem.   The data represent the weight gain of two
groups of children over a six month period.   The first sample consists of 10
observations of weight gain under diet 1 over a sixth month period, the
second has 9 observations of weight gain under diet 2.   The data is stored
in two column vectors called C1 and C2.
Summary statistics are easily obtained using the describe command with the
following result.

```
MTB > DESCRIBE C1 C2

    N   MEAN   MEDIAN  TRMEAN  STDEV  SEMEAN  MIN    MAX     Q1     Q3

C1 10  11.570  11.650  11.512  1.520  0.481  9.60   14.00  10.10  12.85

C2  9  16.47   15.30   16.47   3.98   1.33   11.00  22.00  13.00  20.35
```

We are presented with a wide range of summary statistics which need to be understood to be used properly.   They include all the traditional measures but also a robust measured, the 5% trimmed mean, which therefore should perhaps be briefly described as part of the course.   This measure plays a useful role of validating the mean particularly when used in conjunction with the median.   Between them they give us information on the symmetry and possible presence of an outlier.   Small differences between the three measures increase our confidence in using the mean in subsequent calculations.   The ease with which these values can be obtained should encourage us to describe how to interpret them.   In this way the output produced by the package can impinge beneficially on the content of a course.

Histograms of data are readily available using a single command. However, particularly in the computer age, they should not be the sole visual summary that is taught.   Useful plots such as the Box-Plot (Tukey [11]) allow for simple data summary and have greater visual impact than the usual histogram.   They can be used with much smaller samples, allow much easier comparison of several samples and are much more helpful in encouraging the students to examine their data before they proceed with the analysis.   The box-plots for the two samples described above are shown below.

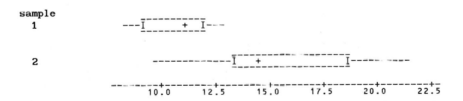

We can see from this the relative location and variability in the two samples.

A two sample test between the means can be carried out by the command TWOSAMPLE C1 C2 which produces

```
TWOSAMPLE  T  FOR  C1  Vs C2

            N      MEAN      STDEV      SEMEAN
      C1   10     11.57      1.52       0.48
      C2    9     16.47      3.98       1.3

95 PCT C1 FOR  MU C1 - MU C2: (-8.04, -1.8)
TTEST  MU C1 = MU C2 9VS NE): T = -3.47 P=0.0060 DF=10.1
```

The output gives a 95% confidence interval for the difference between the means, the t statistic, p value and the degrees of freedom.   Thus no tables are needed, only knowledge of how to make a decision about the hypothesis under test.   It should however be noted that the degrees of freedom given are non-integer reflecting the approximate test for unequal variances which is being presented, in this case, as the default.   To obtain the test based on the assumption of equal variances, using a pooled estimate of variance, it has to be specifically requested, using a subcommand, as follows

```
MTB >  TWOSAMPLE C1 C2;
SUBC>  POOLED.
```

The resulting output is

```
TWOSAMPLE  T  FOR  C1  Vs  C2
         N     MEAN      STDEV     SE MEAN
   C1   10    11.57      1.52       0.48
   C2    9    16.47      3.98       1.3

95 PCT  C1  FOR MU C1 - MU C2; (-7.75, -2.0)
TTEST MU C1 = MU C2 (Vs NE): T= -3.62 P=0.0021 DF=17.0
```

It is this test which is traditionally taught and assumed in most courses since it has an analytical solution.   The case of unequal variances does not have an analytical solution and the distribution used in the output from the package is only an approximation.   The students therefore need to be clear of the assumptions under which they are carrying out the test, and the situations in which the approximation is valid, a matter of some controversy in statistical circles.   This lends more support to the argument that the ideas of the robustness of statistics should be discussed.   It also illustrates the importance of knowing the underlying principles behind the output that is generated.   Using output of this type and working with a standard statistical text, in which the equal variance case is often the only one described, may also cause some conflict for the student.   Problems such as these need to be anticipated when using a package in a course.

Placing the emphasis on the output from the package rather than how it is being produced allows us to introduce the student to these wider aspects of the problem. Thus in this example we can place greater emphasis on the correct choice of hypothesis, the decision making process and in particular the assumptions on which the decision is based.

3.1 Advanced Courses

The more comprehensive packages SAS and GENSTAT have been used by final year undergraduates and by masters students. Our early courses involving use of a computer were based on the GENSTAT package. It allows a variety of statistical methods to be used although the emphasis is on the linear model. Its operation requires writing short programs and hence is cumbersome when compared with the SAS package that subsequently became available to us. An important aspect of using computers in any course is the ease with which the students can use them and this, together with the wider range of facilities available in SAS prompted the change.

SAS operates using a set of macros which allow a wide-ranging set of statistical and data management operations to be carried out in a relatively easy way. It does however generate a large amount of information, some of it optional, and its use is again dependent on the users understanding what they are doing.

Using packages it is a relatively simple matter to repeat analysis of a set of data removing any non-significant effects from the model. Once the significant aspects of the model have beeen ascertained, ancillary statistics relating to the individual effects are readily available as is diagnostic information relating to the assumptions on which the method of decision is based. Statistical packages therefore make the fitting, validation and modification of models a viable proposition. They also demand a wider knowledge of the statistics that apply to a given problem but at the same time free us from the need to concentrate on calculation techniques. It is also clear from the above that the package can have some influence on the structure of the course. It is however the teachers choice whether to include all the material or advise students to overlook part of the output since it is not discussed in a particular course. It can not be ignored though and may open up new directions for teaching.

4. STATISTICAL EXPERIENCE

From the above we see that problems, as presented in the traditional texts, are completed more easily and in greater detail than was previously possible.

The computer allows us however to add another dimension to the learning process. Rather than present data in the form of a table the problems can be posed in more general terms.   Experimental design problems, for example, can be presented in a way which allows the student to formulate the design. Data can then be simulated from the true model and presented as observations to their design.   Problems posed in this way bear a closer resemblence to a working environment and give students more realistic experience in practising statistics.

A case study approach can, for example, make problems involving response surfaces using techniques such as determination of optimum conditions and evolutionary operation both more interesting and feasible.   These are methods used in industry in areas such as plant evaluation.   They are calculation intensive and, while possible using a calculator, become more interesting using the speed of a computer with which it is feasible to expect students to follow all stages of the design and analysis process.   At each stage a simulated response for each trial in the design can be made available making the work more realistic.   Each student's results can be different and will almost certainly result in different paths being taken to the solution.   By concentrating on the methodology rather than the mechanics of the design, the practice of using such procedures can also be introduced to non-mathematics students such as engineers.

Some packages also include an element of advice about possible problems with the data or the model.   MINITAB, for example, indicates observations in a regression analysis that have large residuals, points exerting high levels of x influence and high levels of correlation between dependent variables. These are very useful guides when fitting models but the user needs to know how to interpret this information and when to ignore it.   It also illustrates another aspect of the computer's potential in the teaching environment.

4.1 Expert Systems

The development of expert systems has a potential in training and guiding students through aspects such as model choice and model validation. Techniques of experimental design and consultancy can also benefit from such approaches.

At Leeds we have developed a FORTRAN based package (TSAT) to assist students to
perform model selection from within the class of ARIMA models (Autoregressive Integrated Moving Average models (Box and Jenkins [4]) that may be used to model time series.   These models can be classified by the correlation structure over different time lags.   Once a model is identified, initial estimates of the model's parameters can be obtained and then estimated efficiently using non-linear estimation routines.   The residuals from the

fit can then be used to validate the assumptions made about the model. This work can all be done within the major packages such as SAS, GENSTAT, BMDP and SPSS, but in each case there is considerable work to be done in setting up the various routines and analysing the output at the various stages. While this is adequate for working on a single problem, for the beginner it is time consuming and does not really give them the opportunity to gain the experience that comes from analysing many series.

The package is designed to be used in two parts. The first gives practice in model selection. Students are presented with a graph of the time series and the necessary identification statistics. They are then asked to make a choice of model using this output. This is then entered into the package and its plausibility checked. If it is not a reasonable model, the package suggests an alternative giving an explanation for this choice, allowing the student to compare his reasoning with that of the 'expert'. Figure 1 illustrates part of a simple session using the package Students have commented how useful they find this and very soon we find that they improve their techniques of model choice. It also helps to give them experience, since in a one hour period they can work through about 10 series rather than one or two. This allows repetition of cases to be handled and a greater variety of examples to be presented. The package can also be used to fit both the user's chosen model and that chosen by the package, if different. The various diagnostics can then be examined together with comments on the adequacy of the models. The student is asked to suggest any modifications that might need to be made to the model and the package comments on them. This work can all be done on line and the revised models fitted. The cycle is then continued until no more changes are necessary to the model and the data is adequately described, as far as is possible within the limitations of the class of models.

The advantage of working in this way is that the need to program a package is removed and the students are quickly able to learn from their mistakes. Several time series can be considered in the time that a single time series can be analysed using most statistical packages available. Once the student has practiced applying the methodology they can find the interpretation of output from the classical package much easier to cope with when they come to use it. They are able to reach decisions much quicker and much more accurately when compared to students who took the course before we had the package.

Another area where the practical skills of applying methodology can be greatly enhanced using an expert system is in design of experiments. They give us an opportunity of placing the student in the position in which they

Identification of

Wolfer Sunspots Annual Data

The data has a mean of 46.9 and a variance of 1382.2

The Autocorrelations for lags 1-20 are
0.805 0.426 0.068 -0.171 -0.268 -0.212 -0.043 0.165 0.332 0.412
0.395 0.288 0.142 0.019 -0.054 -0.102 -0.145 -0.177 -0.167 -0.102

Do you think any differencing is necessary? **No**

The partial autocorrelations for lags 1-20 of the undifferenced data
are
0.805 -0.630 0.076 -0.666 0.006 0.169 0.103 0.121 0.074 0.080
0.069 -0.036 0.077 0.043 -0.039 -0.130 -0.155 -0.121 -0.017 0.002

An approximate 95% confidence interval is -0.196 to 0.196

What values do you propose for parameters P and Q? **2 0**
Do you think a constant term should be included in the model? **No**

```
-------------------------------------------------
You have chosen a (2,0,0) model without a constant
-------------------------------------------------
```

The package agrees that there is no differencing required.

With no differencing the package chooses a model as follows

The autocorrelation functions shows evidence of slow decay since
more than 3 of the terms in the first 10 are significant.   The
partial autocorrelation function on the other hand has only the first
two terms significant in the first ten hence an autoregressive model
of order two is the parsimonious choice.

Thus the package agrees with your choice of a (2,0,0) model.

The mean however is greater than 2 standard deviations from zero once
the latter is adjusted by the model chosen hence is significantly
different from zero and should be included in the model.

Do you want an explanation of this point? **NO**

The initial estimate of the constant is 14.9
 while the error variance is 293.00

Figure 1   *Illustration of the display from the TSAT package showing*
how students can be guided through the complexities of choosing models
for time series.

can practice handling questions about the broader aspects of a problem.

Expert systems, such as SELINA (Baines and Clithero [2]) are being developed to assist managers and other non-statisticians (experimenters) in designing experiments. These operate by asking questions related to the formulation of a problem and then negotiate with the user until a suitable choice of design can be presented. These packages also have a potential role in statistics teaching of assisting students in learning the type of information to extract about a problem at the planning stage. They need for example to learn about

1) specifiying the variables used in a design

2) formulating any necessary block structure and

3) specifying the treatment structure.

In more complicated problems they need to make decisions on the effects that are to be confounded. They can achieve this by placing the student in the experimenters or managers position thereby giving them experience in handling the type of questions they as statisticians should be asking. Such an approach can help to give students experience in the techiniques of consultancy which are fuundamental to the practise of statistics.

5. CONCLUSION

The modern student of Statistics needs to know how to handle and analyse data as well as mastering the mathematical theory underlying the methodology.

This can be reflected in our teaching as access to computers enables us to produce a better balance between the theory and application. They can be used to give experience in applying the methods to real data sets, exploring data sets and validating the model chosen and the assumptions made. This enables us to present the students with practical problems that require them to

a) ask relevant questions about real problems

b) relate the statistical models covered in the theory to physical situations

c) present meaningful conclusions drawn from the avaiable situation.

The ability to perform calculations quickly has removed many of the restrictions on both the size of the problem that can be handled and the methods that can be used. It also has an effect on the type of mathematical background required. For example, developments in regression and design of experiments have resulted in the emphasis moving from calculation of sums of squares for well-balanced situations to handling and interpreting the results of matrix algebra, leaving the computer to perform the calculations.

We also need to teach how to interpret a problem and apply the correct methodology.   It can be argued however that it is only the mathematical statistician who needs training in the mathematics behind the theory and the ability to handle and develop methods for non-standard problems.

The user of statistics does not need advanced knowledge about the theory and calculation techniques related to statistical methods.   He rather wants to know how to select the appropriate method for analysis and how to apply and interpret the results obtained by using it.   Statistical packages have made the application easy and a course based on using a statistical package, with examples related to the student's field of study, has more potential benefit than a course based on the classical introductory text.   The teaching can be based on application and illustration, together with practice and personal experience, without having to cope with the mathematical complexities of statistical theory.   This has had an invigorating effect on service teaching through its wider appeal to the students and the greater scope of material covered.

We conclude therefore that the computer is having a variety of influences on the teaching of statistics.   It can assist at the introductory stage in the interpretation of mathematical ideas.   Here it should be used to compliment rather than replace practical statistics.   The triumvariate of mathematical theory, experimental work and computation permits us far greater scope when introducing and explaining the underlying concepts of statistical practice.

Secondly, the computer aids and broadens the amount of statistical calculation that can be expected.   Statistical packages are the simpliest way in which this can be done.   The design of the package, particularly the information it supplies, also encourages and requires careful consideration of the material introduced in a course.   To simply expect students to use a computer to perform the calculations generated by a traditional course would result in missing many of the benefits that can result from using a package. It would also fail to reflect the many changes that are occuring in the subject in the age of the computer.

REFERENCES

[1] Atkinson A.C. Plots Transformations and Regression: An Introduction to Graphical Methods in Diagnostic Regression. (Oxford,1985)
[2] Baines A and Clithero D.T.   Interactive User-Friendly Package for Designing and Analysis of Experiments in: De Antoni F., Lauro N. and Rizzi A. (eds.) Compstat 1986(Springer Verlag)
[3] Bousbaine A.,Kent P and Neusch P. Are Statistical Tables Obsolete? in: Rade L. and Speed T.P. (eds.) Teaching Statistics in the Computer Age

(International Statistical Institute,1986)

[4] Box G.E.P and Jenkins G.M Time Series Analysis Forcasting and Control (Holden Day, 1976)

[5] Cooke R.D. Technometrics 19 (1987) pp 15-18

[6] Diaconis P and Effron B. Scientific American 248 (1983) pp96-108

[7] Green D Teaching Statistics 1 (1979) pp66-70

[8] Holmes P. Using Microcomputers to Extend and Supplement Existing material for Teaching Statistics. in: Rade L. and Speed T.P. (eds.) Teaching Statistics in the Computer Age. (International Statistical Institute,1986)

[9] Huber P.J. Robust Statistics (J.Wiley and Sons,New York,1981)

[10] Speed T.P. Teaching Statistics at University Level in:Rade L. and Speed T.P (eds.) Teaching Statistics in the Computer Age (International Statistical Institute,1986)

[11] Tukey J.W. Exploratory Data Analysis (Addison and Wesley,1977)

[12] Weisberg S. Applied Linear Regression(J.Wiley and Sons,New York,1980)

[13] Yates F. Biometrics 22 (1966) pp233-251

ECM/87 - Educational Computing in Mathematics
T.F. Banchoff et al. (editors)
© Elsevier Science Publishers B.V. (North-Holland), 1988

THE TEACHING OF MATHEMATICS IN A COMPUTER AGE

D. L. SALINGER

School of Mathematics, Leeds University
Leeds LS2 9JT, England

ABSTRACT

We discuss the benefits and problems of introducing computers into
mathematics teaching in the context of the mathematics course at
Leeds University. Particular reference is made to the effect of
the European Community's Joint Study Programme with Paris-Sud and
Rome.

1. INTERRELATION OF MATHEMATICS AND COMPUTING

The aims of a mathematical education are varied and often ill-defined, if
explicitly defined at all. Nonetheless, there is a measure of agreement, even
at the international level, about what should be present at the basic stages:
calculus/analysis, linear algebra/geometry, mechanics. Why? What is taught
is in part what mathematicians think will be useful to engineers, scientists
and teachers in secondary education; it is in part what "no future
mathematician can do without". It is not based on the historical development
of mathematics, though we may tend to teach 17th to 19th century mathematics.
Mathematics is usually taught according to the internal logic of the subject
matter (which imposes its own choice of beginnings); what historical
background there is is usually an aside. Moreover, it is taught with 20th
century set-theoretical background and 20th century rigour. Above all, the
curriculum is determined by what we have most experience of: our knowledge of
contemporary mathematics and research practice.

That practice has changed over the last twenty-five years, with the advent
of digital computers. Fast calculation is not now available only to the
Gausses of our age. I say that to emphasize the continuity; mathematicians
have always undertaken long calculations: Napier spending twenty years over
the compilation of his tables of logarithms [2] or Gauss calculating data on
which to base his conjectures or Euler reputedly ruining his eyesight in a
three day astronomical calculation [6] .

Computers can now take us further, with less danger to health, either in
making deep conjectures (for example, the Birch - Swinnerton-Dyer conjectures)

or settle problems such as Waring's problem, where part of the work is a mammoth calculation.   As is demonstrated elsewhere in this volume [1,8], the computer's graphics capabilities allow us to visualise complicated sets, such as the Julia and Mandelbrot sets, or differential manifolds and hence make further progress in their mathematical description.   More controversially, we have computer-aided proofs - such as that of the four-colour theorem or the classification of finite simple groups.

Computers have given rise to their own disciplines, not all mathematical, but including numerical analysis, computer algebra and the swathe of subjects known as theoretical computer science.   These include automata theory, recursive function theory, complexity theory, denotational and operational semantics,   the   theory   of   correctness   of   programs,   feasibility   and intractibility: they should   begin to find a place in the mathematics curriculum.

The way that mathematics is used in society is changing as well.   Where, before, experts were needed with considerable mathematical knowledge, perhaps supplemented by mathematical tables, there are now computer packages.   These need much less mathematical knowledge to get some sort of result out of, but are   more   powerful   and   their   limitations   are   not   generally   understood. Engineers and others might like to have some universal program which will solve each of their particular problems: such things may not be possible for theoretical reasons.   Students of mathematics need sufficient understanding of these matters to be able to "blow the whistle".   At the same time, as symbolic manipulation gets more efficient and more widely available, there will be less demand for the traditional calculus and matrix manipulation courses.

2.   COMPUTING AS PART OF A PARTICULAR MATHEMATICS CURRICULUM

I shall now take some space to describe the practice of my own mathematics department.

We teach a variety of courses to students, who come to university having already chosen what they will do for the three or four years of their undergraduate career.   Perhaps a third of the ten thousand students at Leeds university take some tuition from the School of Mathematics during their stay. These include physicists, chemists, engineers, economists, computer scientists and a wide variety of arts students.   What I shall describe, however, is the course for those students who opt for the specialist mathematics course.   We recruit about eighty a year of whom between eighty and ninety per cent graduate after three years.

Their first year consists of compulsory courses, the second and third years have compulsory courses in pure and applied mathematics, but there is room for a large choice of options: in the final year there are fifty options from which a student chooses at least eight. The content and number of courses is dependent on decisions made within the School of Mathematics (certain of those decisions have to be approved by a university committee, but such approval is usually a formality).

The framework of the course was agreed upon some fifteen years ago as the result of considerable discussion within the School of Mathematics, which then included the Department of Computer Science (the School is now comprised of the Departments of Pure Mathematics, Applied Mathematical Studies and Statistics). The outline of compulsory courses is settled by individual departments. The content of optional courses is up to the lecturer giving them.

I do not wish to give the impression that the Mathematics degree course is purely the conscious result of the considerations outlined in the first section; courses evolve from earlier courses and the overall shape of the curriculum owes something to administrative considerations.

Among the skills of a modern mathematics student a sine qua non is familiarity with a programming language. For many people that statement might need some justification: could not students use appropriate packages? But in an education - as opposed to training - students need some mastery of the medium of computing and some understanding of its intrinsic limitations. Thus they need to learn a "high-level" language. Which language and on what machine?

One could - and one does - waste time on discussing which language a student should start with, but I am not a purist and don't mind. At Leeds, we have opted to use PASCAL, specially designed for teaching, but BASIC and FORTRAN 77 have their advocates because they are among the most widely used languages. Whatever language is used, it's important that students quickly learn to get the computer to do some basic mathematics. What matters is the experience of programming: finding out what can go wrong and what can go right. So we run an initial programming course in the first term that the students are here. In fact we give them a go on the micro-computers before their mathematics lectures start.

The School of Mathematics is fortunate in possessing a micro-lab and in having access to good multi-user mainframe facilities. The students have free entry to the micro-lab, subject to availability of the machines and a reasonable ration of time on the mainframe if they need it. We have run the initial computing course both on mainframe and micros. The initial stages are far easier on the micros, but apart from that, it's not clear which is ultimately the best choice. One advantage of a mainframe is that one can monitor the use of the machine by each student and thus pick up far earlier

the few students who drop out.   Also useful is the immense power and virtually
unlimited storage facility of such a machine.   Our micros are BBC ACORNS
(Olivetti), which are now somewhat elderly in concept, their chief
disadvantage being a very small memory, but they are quite fast with
excellent colour graphics.   They are linked by a network, which enables a
common printer to be used and files to be shared, but they have individual
disc drives.   The version of PASCAL on the BBCs is ISOPASCAL which has
extensions to access the BASIC graphic routines of the micros.

The initial programming course consists of twelve lectures and twelve
weekly one-and-a-half hour supervised sessions on the machines.   If the
students don't find that enough, they can use the micro-lab unsupervised when
it is not otherwise used.

In the time available, we cannot usefully teach all of PASCAL.   Roughly,
what we cover is the following:-

the standard scalar types;

constant definition;

assignment statement;

while, repeat and for statements;

if statement;

declared scalar type;

subrange type;

array type;

filetype;

procedures and functions;

input and output.

We do not have time to teach record or pointer types, nor can we do more
than mention the possibility of recursion.

Of course, we don't do much on text handling or data-bases: after all, we
must concentrate on mathematics.   The emphasis is on acquiring practical
programming skill: students are required to write programs that run, but we
describe the algorithms to them.   For example, they may be told what the Euler
method is and asked to apply it to a particular differential equation.   They
may get an informal explanation of why the Euler method can work, but no
analysis of how or when it goes wrong.   However, they may be asked to compute
a particular case of its failure.

Examples are drawn from pure and applied mathematics and statistics
related, as far as possible, to the courses the students take.   Grading is by
assessment of four short projects.

The programming course is followed by a numerical analysis course which
covers many of the topics implicitly encountered in the programming section:
zero-finding, polynomial interpolation, numerical integration, numerical

solution of ordinary differential equations and *Gaussian elimination.* Grading is mainly by examination, but a small proportion of the marks is obtained on course work.

This year, we introduced short mathematical projects into our first year curriculum. Several of these - *coding, motion of a particle with air-resistance, bird navigation,* - gave opportunities for students to use computers.

After the first year, courses involving the use of computers are optional, but those with a mathematical computing flavour can include - in the second year of a three year course: *Computational numerical methods, Logic, Theory of Computer Science* - and in the third year: *Computer Solution of Differential Equations, Numerical Methods for Partial Differential Equations, Computability, Theory of Programming Languages, Complexity Theory, Advanced Numerical Analysis.*

Some of these courses are taught by computer scientists, some by mathematicians (and, of course, some of the computer scientists are and regard themselves as also mathematicians; some of the mathematicians find it politic to regard themselves as computer scientists, at any rate when making grant applications.)

3. THE IMPACT OF INTERNATIONAL CO-OPERATION

M. Dechamps [3] describes, elsewhere in this volume, the European Community's Joint Program of Study linking the mathematics departments of Rome, Paris-Sud and Leeds Universities. Here, I would like to outline some of the benefits of a joint effort.

Firstly, in seeing how others have handled introductory computing, we have been able to see how much manpower would be involved. In the initial stages of learning to compute, we have decided, on the basis of experience at Leeds and Paris-Sud, to have a very high teacher-pupil ratio, perhaps 1 to 6. Later on, we reduce that, but, at the start, it's important to preserve students' confidence. In the past, we have experienced an unacceptable number of students failing to come to grips with computing, at least partly due to a lack of friendly help at the start.

The programming course and its examples were and will be influenced by the exchanges of ideas between Rome, Orsay and ourselves. Several of us have had the chance to teach at the co-operating institutions, and have been able to profit from different approaches to the same subject. For instance, I intend to incorporate a graphics library in the initial course next year and have used illustrations from the course at Orsay on the limitations of the computer to great effect with a variety of students and intending students. My own support for project work using computers at Leeds has been greatly influenced by the experiments I have seen at Rome University and Turin Polytechnic.

There are also features of the teaching at Rome and Paris-Sud that we like but have so far not been able to incorporate in our work at Leeds: the longer project and students writing their own teaching package being two examples.

At Leeds we have worked with two groups of French students, on partial differential equations and zero-finding, respectively. For the students, it has been an opportunity to use the mainframe facilities which we can offer, including (for the PDEs) great speed of computation, advanced graphics packages (though the students were determined to write their own 3-d plot routine) and (for the zero finding) a sophisticated colour graphics printer.

Two groups of three Leeds students each have gone to France and have successfully completed projects on polynomial interpolation and PROLOG. They could, if they wished, continue the project started at Paris-Sud into their final year at Leeds. One of the second group has done so. Of the first group, two have been accepted to do post-graduate work at Leeds, one in theoretical computer science and the other in an area which will involve a great deal of practical computing.

For the students it has also been a social experience and has given them a chance to experience a different approach to learning arising out of the intellectual tradition of another European country. These are tentative beginnings though. In the future, we should anticipate students spending whole years in another country and being able to count the work towards their final degree, as the ERASMUS project of the European Community envisages.

4.   COMPUTING AS A TEACHING AID

Is the computer a powerful teaching aid or will it give just the illusion of helping to teach? There are those, perhaps unfamiliar with both computers and teaching, who say things like "university teachers will soon be unnecessary: all teaching can be carried out by computers and videodiscs." In that view our future will be to prepare teaching packages and then to retire!

If we think of a teaching aid as something that helps us to do our existing teaching better, then computers can indeed be useful. I like to use the computer to generate better diagrams than can be managed on a blackboard or hand-drawn projector slide, where it can be done by short programs. For example, I have used a computer in the classroom to illustrate analytic functions, uniform convergence and Fourier series. Elsewhere in this volume, D.G. Knapp [7] describes its use in illustrating solutions of differential equations. These are places where the computer can enhance the imagination and stimulate students to have intuitive pictures of mathematical objects.

So, yes, the computer is useful to generate better diagrams than one can produce with chalk. One can use it as an expert draughtsman or as a moving blackboard. Further, if there is a minimum amount of flexibility in the program, the students can themselves experiment with drawing sequences of functions, solutions of differential equations or whatever. As long as they don't switch off their brains, this can be helpful in stimulating their imaginations.

Preparing short illustrative programs for one's own use is one thing. Writing packages which can be used by other teachers in other places is a much more demanding activity. I shall proceed to describe two teaching packages that are under development at Leeds: one on techniques of integration and the other the LEEDS LOGIC PROGRAM.

The former, which I have been involved in developing, is set up in the context of a general suite of packages in various disciplines, going under the name of FOAL, developed by Kenneth Tait in the Leeds Computer Based Learning Unit [9]. The framework is that of multiple choice questions. Students can test their abilities on questions which are randomly accessed from a question bank. They get a score for each session on the computer (and their progress can be monitored by the class tutor). For the integration package, we have chosen to lead students through harder examples in up to four stages, in each of which they have to choose the correct answer from up to five possibilities. If students make a wrong choice, they get as helpful a short comment as we can think of. They have then to return to making the choices. At the end of the four stages, there is a summary consisting of the questions and the correct responses.

The system is limited in having no graphics and having to work only with standard ASCII characters. On the other hand, the questions are written in an author language so that the programming aspects are all taken care of. The questions can be written on a simple form and then typed into the computer by an assistant, or else typed directly into a file on the computer. I choose to do the latter: it takes me about one hour for each four part question. This gives a better ratio of time taken in preparation to amount of teaching material than the 200 to 1 I have seen put forward [5].

The Leeds Logic System is a joint project of the departments of Mathematics, Philosophy and Computer Studies, funded by a grant from the Science Research Council. From later this year, it will involve St.Andrews University as well.

In its present, provisional, form, it asks students to deduce one statement in Propositional or First Order Predicate Calculus from another using rules of natural deduction. There are quite a few logic teaching systems already in use. What distinguishes the Leeds system is that the rules of the logic system are flexible. They can be added to either by teacher or student. And rules can be removed, as well. The present version (2.5) has been programmed by Antoni Diller [4].

5. THE EFFECT OF COMPUTERS ON WHAT WE TEACH

Suppose now that we engage - as I've argued that we must - in all the frantic activity that introducing computers into the mathematics curriculum involves. We've looked at possible benefits; what can be said about the disadvantages?

The first is that - as with anything new - it takes time and thereby displaces something else. One approach is to offer students a choice of curriculum: one traditional, the other with computers. Another approach is to allow the more able students to do the extra work involved. But, given the choice, most students will want to have some familiarity with computing within their subject, particularly as they think their future employers will expect them to have that familiarity.

Learning programming is very time-consuming. Initially, it needs a large input of teaching power (in terms of time and people, not intellectual effort); it can absorb more of a student's work time than an objective observer might judge worthwhile. It has been a frequent complaint of our students (and of some of our members of staff) that computing takes up time that could be better spent on learning mathematics.

There are always those students, too, who disappear into the machine, as it were, becoming mesmerised by the screen. Some of those can switch to studying computer science instead of mathematics, but they don't necessarily do well even then. Making a computer do things can be so much easier than the intellectual effort of mastering an academic discipline that students can fool themselves that they are making progress when they are just marking time.

Then, some things in mathematics are more capable than others of being aided by computers. Subjects with a strong visual element, such as geometry, topology and differential equations are particularly suitable. Will this tend to change the balance of what we teach? Will aesthetic considerations outweigh academic arguments in the design of the curriculum?

Computers can give totally incorrect results: because of mistakes in the programs, for instance, or because of rounding and trucation errors, or even because of overflows being automatically set to zero. Can we carefully control what students see of computers and do with computers, or do we teach them the extra numerical analysis needed to encourage a proper scepticism about numerical results? I incline to the latter view, but that, too, takes time and is often not a popular option with the students themselves.

How much should we replace the teaching of routine symbolic manipulation and numerical calculation by computer packages? There is a problem here for weaker students. To some extent, they can survive in a mathematics course by concentrating on lengthy calculations rather than on abstract mathematical constructs. If computers remove the need for the former, what can such students do? And will students be dependent on having a sufficiently powerful machine at hand to cope with even the simpler examples?

We are also faced with a dilemma, because we want students to be able to handle the *concepts* of calculus: will they be able to do so when the calculations are performed not by the students themselves, but by MACSYMA or MUMATH or REDUCE? Further, the traditional methods of integration, for example, are used in theoretical applications in mathematical theory: we need our students to know those methods, even if the computers solve things by other means. Again, we might want students to continue to be able to multiply matrices or evaluate determinants because of later theoretical uses:- I can think of examples in operator algebras, ring theory, K-theory and the geometry of Banach spaces as well as in linear algebra itself.

Some of the drawbacks of using computers can be avoided if they are borne in mind by the teacher in a particular situation. It's clear that teaching packages can't be guaranteed against bad use, at any rate if they're flexible enough to be used creatively by pupils or teachers. There will be bad teaching with computers as well as good. Let us hope, however, that the enthusiasm generated by computers – and so much in evidence at this conference – will compensate for any faults and help convince students that mathematics is still alive and exciting.

REFERENCES

[1] Banchoff, T., EDGE: the Educational Differential Geometry Environment, in this volume.
[2] Baron, M.E., Napier, John, in: Dictionary of Scientific Biography (Charles Scribner's Sons, New York, 1974).
[3] Dechamps, M., A Joint European Effort in the use of Computers in Mathematics, in this volume.
[4] Diller, A., Thrills and Illative Combinatory Logic, in: Proceedings of Workshop on Computer Logic Systems (University of Leeds Centre for Theoretical Computer Science, in press).
[5] Elliott, R., Computer Facilities for Teaching in Universities, Report of the Working Party of the Computer Board (1983).
[6] Fuss, N., Lobrede, in: Leonhardi Euleri Opera Omnia Series 1 Vol. 1 (B.G.Teubner, Leipzig and Berlin, 1911) pp. XLIII-XCV.
[7] Knapp, D.G., Dynamical Systems, in this volume.
[8] Peitgen, H.O., The Beauty of Fractals, in this volume.
[9] Tait,K., The Building of a Computer-Based Learning System, Comput. Educ. 8 (1984) 15-19.

3. SOFTWARE REPORTS

Name of
contributor: I.CAPUZZO DOLCETTA   M.FALCONE

Institution: DIPARTIMENTO DI MATEMATICA, UNIVERSITA' DI ROMA "LA SAPIENZA"

Address: PIAZZALE A.MORO,2   I-00185 ROMA – ITALIA   (06) 4763213

Title of software package: CALCULUS

Description of what software does [200 words]:

CALCULUS is a package designed as an informatic support in teaching a calculus class. The general philosophy underlying this software is that the numerical and graphical facilities of a personal computer can improve the student's understanding of analytical ideas backing a theoretical approach to some specific problem and, on the other hand, substantially contribute to a creative and critical approach to mathematics.
The CALCULUS menu includes the following highly interactive programs:

1.   Computational errors
2.   Sequences of numbers
3.   Numerical series
4.   Graphic representation of real functions
5.   Parametric plane curves
6.   Taylor expansion
7.   Numerical methods for solving $f(x) = 0$
8.   Computing definite integrals
10. Systems of linear equations
11. Surface drawing
12. The Cauchy problem
13. 2X2 systems of linear ODE's

Potential users: UNDERGRADUATE CALCULUS STUDENTS

Fields of interest: SEE ABOVE

It Is: [x]Application program   [ ]Utility   [ ]Other_____

Specific area <u>CALCULUS,DIFFERENTIAL EQUATIONS</u>

Software developed for [name of computer(s)] <u>IBM PC AND COMPATIBLES</u>

in [language(s)] <u>BASIC,PASCAL</u>

to run under [operating system] <u>MS DOS</u>

and is available in the following media:

[x]Floppy disk/diskette. Specify:

    Size <u>5 1/4 IN.</u> Density <u>DD</u>_____   [ ]Single-sided   [^X]Dual-sided

[ ]Magnetic tape. Specify:

    Size _____   Density _____   Character set _____

Distributed by: <u>N.ZANICHELLI EDITORE,VIA IRNERIO 2,   BOLOGNA—ITALIA (*)</u>

Minimum hardware configuration required: <u>IBM PC WITH 2 DISK DRIVERS AND CGA</u>

Required memory: <u>256 K</u>   User training required: [ ]Yes [^X]No

Documentation: [ ]None [ ]Minimal [x]Self-documenting
               [ ]Extensive external documentation

Source code available: [ ]Yes   [ ]No

Level of development: [ ]Design complete   [ ]Coding complete
  [x]Fully operational   [ ]Collaboration would be welcomed

Is software being used currently? [x]Yes   [ ]No
  If yes, how long? <u>2 YEARS</u>   If yes, how many sites? <u>UNIVERSITA'DI ROMA</u>

Contributor is available for user inquiries: [x]Yes   [ ]No

THE SOFTWARE "CALCULUS" IS DISTRIBUTED IN CONJONCTION WITH THE BOOK BY THE SAME

AUTHORS "L'ANALISI AL CALCOLATORE"

Name of
contributor: _____ Prof.Harley Flanders _____

Institution: _____ University of Michigan _____

Address: _____ Department of Mathematics University of Michigan ____

            Ann Arbor, MI 48109   USA _____

Title of software package: _____ MicroCalc 3.0 _____

_____

_____

Description of what software does [200 words]:

MicroCalc provides an interactive laboratory environment for teaching calculus,with two and three dimensional graphics,symbolic
manipulation,and numerical methods.

There are 34 modules and three utilities,covering almost all
topics taught in calculus courses.

User input is similar to the way mathematics is written,e.g.,

                    2 sin 3x cos 4y

Functions that are input or output of one module may be used in
other modules.

Potential users: _____ Calculus instructors and students _____

Fields of interest: _____ Mathematics laboratory software specific to calculus

It Is: [ ]Application program   [ ]Utility   [ ]Other_____
Specific area _____Calculus_____

Software developed for [name of computer(s)] _PC/XT/AT and Apple IIe/IIc_
in [language(s)] _____Pascal_____
to run under [operating system] _MS-DOS/PC-DOS for PC/XT/AT;p-System_
and is available in the following media:
[ ]Floppy disk/diskette. Specify:
    Size _____ Density _____   [ ]Single-sided   [ ]Dual-sided
[ ]Magnetic tape. Specify:
    Size _____ Density _____   Character set _____

Distributed by:_____MathCalcEduc,1449 Covington Dr.,Ann Arbor,MI 48103

Minimum hardware configuration required: PC/XT/AT with CGA,HGC,EGA;Apple IIe

Required memory: _192 k;128 k_   User training required: [ ]Yes [X]No

Documentation: [ ]None [ ]Minimal [X]Self-documenting
                [ ]Extensive external documentation

Source code available: [ ]Yes   [X]No

Level of development: [ ]Design complete   [ ]Coding complete
  [X]Fully operational   [ ]Collaboration would be welcomed

Is software being used currently? [X]Yes   [ ]No
  If yes, how long? ___3 years___   If yes, how many sites? _____150_____

Contributor is available for user inquiries: [X]Yes   [ ]No

Name of
contributor: ___S.Feliziani V.Franchina C.Mengoni A.Polzonetti___

Institution: ___University of Camerino/CIC - Institute of Mathematics___

Address: ___Camerino (MC) zip code 62032 - Italy___

Title of software package: ___Analog Simulator___

Description of what software does [200 words]:

Program is interactive and simulates an analog computer environment. Traditional implementation steps are reproduced one-to-one with simple commands. Additional facilities include load and store of complete debugged simulations, plot with multi-trace against time and X-Y modes, and interfaces with additional modules. One of these allows a fully automatic generation of the simulation "wiring" schematic, starting from the differential equations set, written in a suitable form.

Potential users: ___Analog Designers, Electronic & Mechanical Designers___

Fields of interest: ___Teaching, Analog Simulation___

It is: [x]Application program   [ ]Utility   [ ]Other _____

Specific area ___Analog Simulation_____

Software developed for [name of computer(s)] _Any UNIX (TM) or MS-DOS machine_
in [language(s)] ___Pascal (Turbo Pascal for MS-DOS)_____
to run under [operating system] ___UNIX (TM) or MS-DOS_____
and is available in the following media:
[x] Floppy disk/diskette. Specify:
       Size ___5"___ Density _DD or HD_ [ ]Single-sided [x]Dual-sided
[ ]Magnetic tape. Specify:
       Size _____ Density _____   Character set _____

Distribuited by: __University of Camerino, Computing Center_____

Minimum hardware configuration required: ___IBM PC/XT_____

Required memory: ___256 Kb (run)___ User training required: [ ]Yes [x]No

Documentation: [ ]None [ ]Minimal [x]Self-documenting
               [ ]Extensive external documentation

Source code available: [x]Yes [ ]No

Level of development: [ ]Design complete [ ]Coding complete
   [x]Fully operational [ ]Collaboration would be welcomed

Is software being used currently? [x]Yes [ ]No
   If yes, how long? ___2 ys___ If yes, how many sites? ___3/4___

Contributor is available for users inquiries: [x]Yes [ ]No

Name of contributor: Pierre JARRAUD

Institution: Université Pierre et Marie Curie

Address: Institut de Mathématiques Pures et Appliquées

45-46, 5ème étage, 2 place Jussieu 75252 PARIS CEDEX 05 - FRANCE

Title of software package:

Basic, Riemann, Darboux

Illustration de l'intégrale sur un micro-ordinateur

Description of what software does [200 words]:

The main purpose is for didactic use :

to illustrate the theory of Riemann integration as taught in the first year of university studies in mathematic,

to provide the students with a more concrete approach of the problems of approximation and convergence related to the existence and the calculation of an integral.

The software represents graphically, for user-chosen functions and bounds, the different ways to approximate the definite integral of a function on an interval of the real line : approximation by rectangles, Darboux sums, Riemann sums, random estimates, trapeziums or Simpson methods. The graph of the function is plotted and the surface to evaluate is lighted on the screen. The corresponding values are displayed in a window.

Windows and menus enable to run the software without programmation. Functions are freely user-chosen as string of characters thanks to a parser of R. Rolland (CIRM Marseille). Syntax rules are called back on the screen.

Two ways of use are possible : either the teacher uses the software during the lecture to illustrate it or the students use it themselves during exercises sessions on micro-computers. In this case, we provide them with an "exercise sheet" to help and guide them by asking questions.

Potential users: First years of University studies in Mathematics

Fields of interest: Calculus (Riemann Integration) Didactic

It Is: [X]Application program   [ ]Utility   [ ]Other_____

Specific area _Teaching of calculus_____

Software developed for [name of computer(s)] _IBM-PC with graphic display____
in [language(s)] _Borland's Turbo-Pascal_____
to run under [operating system] _MS - DOS_____
and is available in the following media:
[1]Floppy disk/diskette. Specify:
    Size _5 1/4___ Density _360 K.____   [ ]Single-sided   [X]Dual-sided
[ ]Magnetic tape. Specify:
    Size _____ Density _____   Character set _____

Distributed by: Université Paris 7, Irem Paris-Sud
                55-56, 3ème étage, 2 place Jussieu 75251 PARIS CEDEX 05
Minimum hardware configuration required: _1 floppy drive_____

Required memory: __256 K_____   User training required: [ ]Yes [X]No

Documentation: [ ]None [ ]Minimal [X]Self-documenting
                [X]Extensive external documentation

Source code available: [X]Yes   [ ]No

Level of development: [ ]Design complete   [ ]Coding complete
  [X]Fully operational   [ ]Collaboration would be welcomed

Is software being used currently? [X]Yes   [ ]No
  If yes, how long? __2 years_____   If yes, how many sites? __2_____

Contributor is available for user inquiries: [X]Yes   [ ]No

Examples of screens

Fonction : t*sin(t*t)

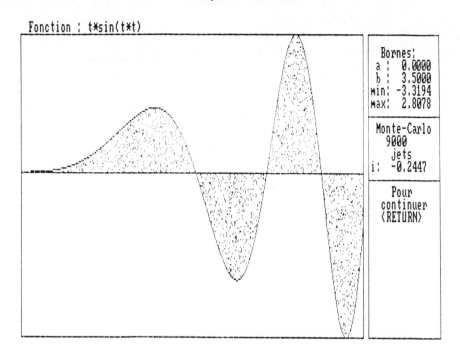

Fonction : 2*frac(t*t/2)

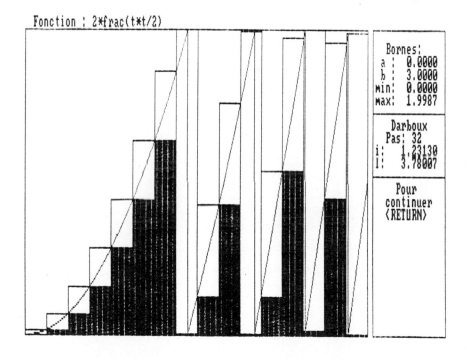

Name of
contributor: ___Hüseyin KOCAK_____

Institution: ___University of Miami_____

Address: ___Department of Mathematics and Computer Science_____

___Coral Gables, Florida 33124   U.S.A._____

Title of software package: ___PHASER: An animator/simulator for dynamical_____

_____systems_____

_____

Description of what software does [200 words]:

PHASER is an interactive graphics-oriented program designed to help create, manipulate, and store various features of ordinary differential and difference equations. For further details on PHASER, please consult author's invited paper in this volume.

Potential users: ___Students and professors   ._____   _____

Fields of interest: ___Ordinary differential and difference equations_____

It Is: [x] Application program   [ ] Utility   [ ] Other _____

Specific area: Ordinary differential and difference equations

Software developed for [name of computer(s)] IBM personal computers _____

in [language(s)] _____

to run under [operating system] __DOS_____

and is available in the following media:

[x] Floppy disk/diskette. Specify:

    Size _360K_   Density _low_   [ ] Single-sided   [x] Dual-sided

[ ] Magnetic tape. Specify:

    Size _____   Density _____   Character set _____

Distributed by: ___Springer-Verlag_____

Minimum hardware configuration required: _256K RAM and CGA graphics_____

Required memory: _256K_   User training required: [ ] Yes [x] No

Documentation: [ ] None [ ] Minimal [ ] Self-documenting
                [x] Extensive external documentation

Source code available: [ ] Yes   [x] No

Level of development: [ ] Design complete   [ ] Code complete
        [x] Fully operational [ ] Collaboration would be welcomed

Is software being used currently? [x] Yes   [ ] No
        If Yes, how long? _3 years_ If yes, how many sites? _many_

Contributor is available for user inquiries: [x] Yes   [ ] No

Name of
contributor: Hervé Lehning and Robert Rolland

Institution: ALE Sup

Address: CIRM   70 route Léon Lachamp

13288 Marseille Cedex 9 FRANCE

Title of software package: MODULOG

Description of what software does [200 words]:

This software is a system of components which greatly simplifies the programming (in Turbo Pascal) of small scientific calculations. This library is divided in two parts :

- the basic kernel which allows one to manage the inputs and outputs; in particular, it contains a graphic kernel admitting all usual cards and a component of the input of numbers, functions and matrices (in numerous and easy ways).

- components which allow one to do a lot of usual computations as integration, differentiation, the resolution of numerical or differential equations and systems of equations.

With some files designed to be included in programs, one can use some types, functions and procedures such as :

Types : Numerical_function, Matrix and Vector.

Procedures : Input_real, Input_function, Input_matrix, etc.

For an example of use, see paper : *Mathematics in a computer age.*

Potential users: Students, software developpers

Fields of interest: Mathematics

It Is: [ ]Application program   [ ]Utility   [X]Other components

Specific area Scientific calculations

Software developed for [name of computer(s)] PC, Mac Intosh

in [language(s)] Turbo Pascal

to run under [operating system] All

and is available in the following media:

[X]Floppy disk/diskette. Specify:

    Size 5"1/4   Density 360 Ko   [ ]Single-sided   [x]Dual-sided

[ ]Magnetic tape. Specify:

    Size _____ Density _____   Character set _____

Distributed by: IREM 70 route Léon Lachamp 13288 Marseille FRANCE

Minimum hardware configuration required: ⁄

Required memory: 64 Ko   User training required: [X]Yes   [ ]No

Documentation: [ ]None [ ]Minimal [ ]Self-documenting
                 [X]Extensive external documentation

Source code available: [X]Yes   [ ]No

Level of development: [X]Design complete   [ ]Coding complete
  [X]Fully operational   [ ]Collaboration would be welcomed

Is software being used currently? [X]Yes   [ ]No
  If yes, how long? 2 years   If yes, how many sites? about 500

Contributor is available for user inquiries: [X]Yes   [ ]No

Name of
contributor: Pascal Laurent, Hervé Lehning and Denis Trystram

Institution: Atelier Logiciel de l'Ecole Centrale (ALE Sup)

Address: Grande Voie des Vignes

92295 Châtenay-Malabry Cedex FRANCE

Title of software package: MATHEMATICAL TOOLS OF CENTRALE

Description of what software does [200 words]:

The mathematical tools of Centrale are designed to allow calculations and simulations on mathematical objects, they are always easy to use. Here is a list of the tools which are available now : *Etudes graphiques*, a curve sketcher, *Etudes des suites et séries numériques*, an assistant to study sequences through checking and testing several conjectures, *Calcul matriciel*, a tool which allows one to achieve computation on matrices, *Etude des surfaces*, a tool which visualizes surfaces. The tools in development are : *Calcul formel*, a tool in formal calculation, *Equations différentielles*, an assistant in the study of differential equations, *Constructions géométriques*, a tool which allows one to achieve geometrical constructions : it will be a nice tool to study loci, *Approximations*, a tool to study approximation.
The way of using these tools is described in the paper : *Mathematics in a computer age*.

Potential users: Students

Fields of interest: General

It Is: [　]Application program　　[x]Utility　　[　]Other_____

Specific area <u>General</u>_____

Software developed for [name of computer(s)] <u>PC</u>_____
in [language(s)] <u>Turbo Pascal</u>_____
to run under [operating system] <u>DOS</u>_____
and is available in the following media:
[x]Floppy disk/diskette.　Specify:
　　Size <u>5"1/4</u>　　Density <u>360 Ko</u>　　[　]Single-sided　[x]Dual-sided
[　]Magnetic tape.　Specify:
　　Size_____　　Density_____　　Character set_____

Distributed by: <u>FIL 36 avenue Gallieni 93175 BAGNOLET FRANCE</u>_____

Minimum hardware configuration required: <u>CGA Card</u>_____

Required memory: <u>384 Ko</u>_____　　User training required: [　]Yes [x]No

Documentation:　[　]None　[　]Minimal　[x]Self-documenting
　　　　　　　　[　]Extensive external documentation

Source code available:　[　]Yes　[x]No

Level of development: [x]Design complete　　[x]Coding complete
　[x]Fully operational　　[　]Collaboration would be welcomed

Is software being used currently? [x]Yes　　[　]No
　If yes, how long? <u>1 year</u>_____　If yes, how many sites? <u>about 50</u>___

Contributor is available for user inquiries:　[x]Yes　　[　]No

FINITE FIELDS ON THE COMPUTER *

Consolato PELLEGRINO

Dipartimento di Matematica
Università di Modena
via Campi 213/B, 41100 Modena, Italy

To write or to use software on a specific subject helps in becoming familiar with the problems involved and it allows a more concrete understanding of their importance. The study of Algebra too can be enriched by such activities and, contrary to what one might ingenuously believe, we can use BASIC. To read or to write software in this (or similar) well known language requires a skill which almost all undergraduate Math students already possess or can get with a very small effort. With this aim we realized the software GF, using known results [1].

GF is a BASIC software that gives the addition and multiplication tables of a Galois field $F=GF(p^k)$, for p a prime and k an integer greater than one. From this software it is easy to write other software which:

- searches for the primitive polynomials over the prime field $GF(p)$;
- allows computations in $GF(p^k)$;
- constructs finite affine or projective planes ([2],[3]), error-correcting codes [4].

REFERENCES

[1] Lidl, R., Niederreiter, H., Finite Fields [Encyclopedia of Mathematics and its Applications] (Addison-Wesley Publishing Co., London, 1983), pp. 89-90.
[2] Dembowski, P., Finite Geometries (Springer-Verlag, Berlin, 1968).
[3] Denes, J., Keedwell, A.D., Latin Squares and their Applications (Academic Press, New York, 1974).
[4] Mac Williams, F.J., Sloane, N.J.A., The Theory of Error-Correcting Codes (North-Holland, Amsterdam, 1978).

* Work supported by CNR.

Name of
contributor: ___Consolato PELLEGRINO___

Institution: ___Dipartimento di Matematica - Università di Modena___

Address: _____via Campi 213/B_____

_____41100 MODENA (Italy)_____

Title of software package: ___GF (Galois Fields)___

_____

_____

Description of what software does [200 words]:

GF is a BASIC software that gives the addition and multiplication tables of a finite field $F=GF(p^k)$. More precisely GF:

1) asks for the prime number p and its exponent k,

2) looks for a primitive polynomial $f(x)$ over the prime field $GF(p)$ and constructs the functions index and exponential,

3) computes the sum of two field-elements,

4) computes the product of two field-elements,

5) prints the field operation tables.

Potential users: ___Undergraduate Math students___

Fields of interest: ___Algebra___

It Is: [x]Application program   [ ]Utility   [ ]Other_____

Specific area ____Algebra_____

Software developed for [name of computer(s)] __CM-64, PC-IBM_____
in [language(s)] __BASIC standard_____
to run under [operating system] _____
and is available in the following media:
[x]Floppy disk/diskette. Specify:
    Size __5" 1/4__ Density __single___   [x]Single-sided   [x]Dual-sided
[x]Magnetic tape. Specify:
    Size _____ Density _____   Character set _____

Distributed by:___author_____

Minimum hardware configuration required: ___base configuration_____

Required memory: __base memory___   User training required: [ ]Yes [x]No

Documentation: [ ]None [ ]Minimal [ ]Self-documenting
               [x]Extensive external documentation

Source code available: [x]Yes   [ ]No

Level of development: [ ]Design complete   [ ]Coding complete
   [x]Fully operational   [x]Collaboration would be welcomed

Is software being used currently? [ ]Yes   [ ]No
   If yes, how long? _one year_____   If yes, how many sites? __five_____

Contributor is available for user inquiries: [x]Yes   [ ]No

Name of
contributor: _____ Fred Simons _____

Institution: _____ Department of Mathematics and Computing Science _____

Address: _____ Eindhoven University of Technology _____

_____ P.O. Box 513, 5600 MB Eindhoven, The Netherlands _____

Title of software package: _____ Calculus with a PC _____

_____ McGraw-Hill Book Company, Hamburg 1987 _____

_____ ISBN 3-89028-207-5 _____

Description of what software does [200 words]:

The diskette contains ten programs to be used in a first year Calculus
course. It accompanies the book "Calculus with a PC" by the same author.
(McGraw-Hill, Hamburg 1987, ISBN 3-89028-026-9) and is published
separately.

The programs ONEVAR and CURVE allow the user to plot the graph of a function
or a curve and to approximate the zeros, the extreme values and the points
of intersection with the axes. The program DIFFEQ plots the direction field
and a solution of a differential equation; the program NUMINT deals with
numerical integration. For analyzing sequences and series the programs
SEQUENCE and SERIES can be used, and there is a program SUCSUB for iterative
sequences and a program ZERO containing some methods for finding a zero
of a function. Finally, the programs TAYLOR and FOURIER demonstrate the
Taylor- and Fourier series.

Potential users: _____ students and teachers of first year Calculus courses. ___

Fields of interest: _____

It Is: [ ]Application program   [ ]Utility   [x]Other_____
Specific area ___Calculus_____

Software developed for [name of computer(s)] ___IBM-compatible PC_____
in [language(s)] ___BASIC_____
to run under [operating system] _____
and is available in the following media:
[X]Floppy disk/diskette. Specify:
      Size _5 1/4 inch_ Density __double___   [ ]Single-sided   [X]Dual-sided
[ ]Magnetic tape. Specify:
      Size _____ Density _____   Character set _____

Distributed by:___McGraw-Hill Hamburg_____

Minimum hardware configuration required: ___IBM-compatible PC_____

Required memory: __64 kB_____   User training required: [ ]Yes [X]No

Documentation: [ ]None [ ]Minimal [x]Self-documenting
                [ ]Extensive external documentation

Source code available: [x]Yes   [ ]No

Level of development: [ ]Design complete   [ ]Coding complete
   [x]Fully operational   [ ]Collaboration would be welcomed

Is software being used currently? [X]Yes   [ ]No
   If yes, how long? ___3 years_____ If yes, how many sites? __1_____

Contributor is available for user inquiries: [X]Yes   [ ]No

4. APPENDIX: Film Festival

After the end of the regular sections and the software demonstrations of the ECM/87,during the evening,a film festival was organized on the theme of the use of computer graphics.

Two films were presented by Thomas Banchoff and his colleagues at the Brown University in Providence:

1) HYPERCUBE: PROJECTIONS AND SLICING.
Produced by T.Banchoff and C.Strauss in 1978 in the film the Hypercube appears first as a square which rotates about axes to appear as a 3-cube then about planes in 4-space to show the full structure of the 4-cube. The film is all in computer graphics animation;duration is 12 minutes;standard 16 mm..*

2) SPHERE AND HYPERSPHERE.
Produced by T.Banchoff,F.Bisshopp,H.Koçak,D.Laidlaw and D.Margolis in 1987.The film is computer-generated and it shows first the stereographic projections of the two-sphere S^2 while the two-sphere is rotated in R^3 ,then the analogous stereographic images of the three-sphere S^3 as the three-sphere is rotated in R^4 .Duration is 3 minutes;standard 16 mm..

Two films were presented by M.Emmer:

3) COMPUTERS.
A films in the series 'Art and Mathematics',produced in 1987, showing how computers have changed the way of working of the mathematicians,the architects,the artists and all scientists. Duration is 27 minutes;standard 16 mm..

4) FLATLAND.
The film,produced in 1987,is the story of the square as it was written by E.A.Abbott in 1884.All the film is in animation with real objects except the last two minutes in computer graphics

---

* Two frames from the film are shown on page 281.

animation   realized   by   T.Banchoff   and his colleagues. Duration is 22 minutes;standard 16 mm..

Two films were presented by O.Peitgen.

5) FLY LORENZ.

The film was produced by H.Jurgens and O.Peitgen in 1984.It is a film in computer animation on the Lorenz attractor and the Chaos Theory. Duration is 13 minutes;standard 16 mm..

6) ZOOMING ON FRACTALS.

The film was produced by H.O.Peitgen, H. Jurgens and D.Saupe in 1986. It describes in computer graphics animation a fantastic zooming on the Mandelbrot set. Duration is 3 minutes;standard 16 mm..

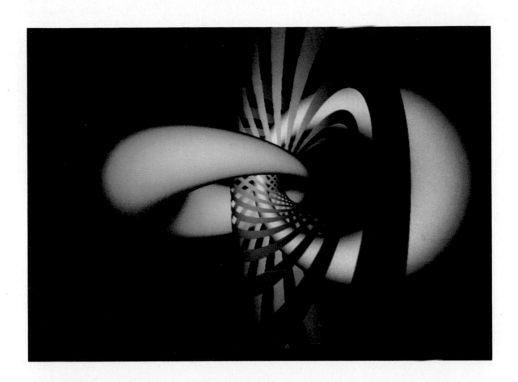

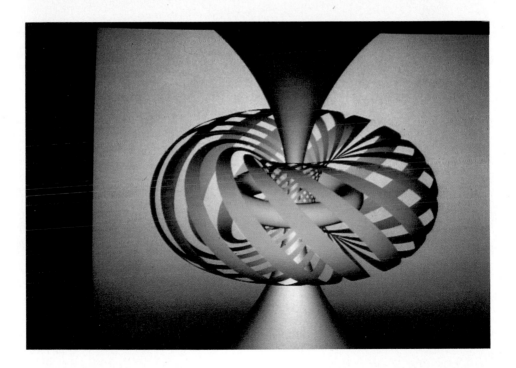

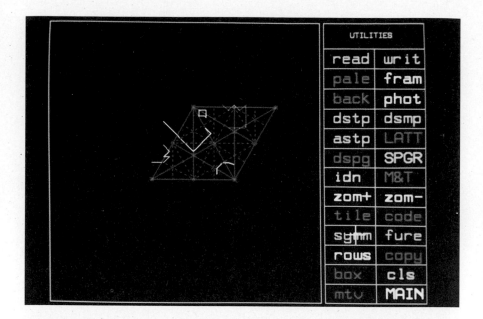

Symmetry elements displayed by Colour Graphics.

(A) — graphic representation of the symmetry elements in the space group p6m: solid and broken lines are mirror and glide-mirror lines respectively; stars and rhombuses are rotation points of order 6 and 2 respectively; rotation points of order 3 are at the intersection of three, and only three, solid lines. The symmetry group acting on the motifs drawn in colour produces the whole pattern shown in (B). (On the right side appears the menu, resident on the video-screen, to call the interactive functions.)

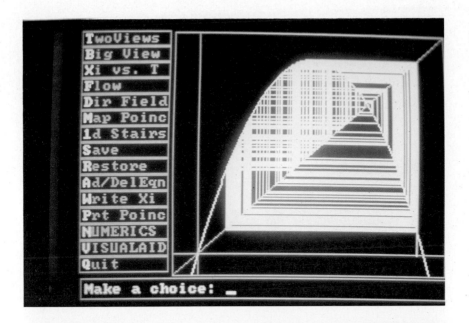

From "PHASER" by H. Koçak

Figure 7

Figure 9

Figure 6

Figure 8

Figures 6, 7, 8 and 9: Similarity between Mandelbrot set and Julia sets